Information Systems Engineering and Management

Volume 63

The book series "Information Systems Engineering and Management" (ISEM) publishes innovative and original works in the various areas of planning, development, implementation, and management of information systems and technologies by enterprises, citizens, and society for the improvement of the socio-economic environment.

The series is multidisciplinary, focusing on technological, organizational, and social domains of information systems engineering and management. Manuscripts published in this book series focus on relevant problems and research in the planning, analysis, design, implementation, exploration, and management of all types of information systems and technologies. The series contains monographs, lecture notes, edited volumes, pedagogical and technical books as well as proceedings volumes.

Some topics/keywords to be considered in the ISEM book series are, but not limited to: Information Systems Planning; Information Systems Development; Exploration of Information Systems; Management of Information Systems; Blockchain Technology; Cloud Computing; Artificial Intelligence (AI) and Machine Learning; Big Data Analytics; Multimedia Systems; Computer Networks, Mobility and Pervasive Systems; IT Security, Ethics and Privacy; Cybersecurity; Digital Platforms and Services; Requirements Engineering; Software Engineering; Process and Knowledge Engineering; Security and Privacy Engineering, Autonomous Robotics; Human-Computer Interaction; Marketing and Information; Tourism and Information; Finance and Value; Decisions and Risk; Innovation and Projects; Strategy and People.

Indexed by Google Scholar. All books published in the series are submitted for consideration in the Web of Science.

For book or proceedings proposals please contact Alvaro Rocha (amrrocha@gmail.com).

Philip Eappen · Narasimha Rao Vajjhala ·
Dimitrios Zikos · Karen Parker Davidson
Editors

Remote Monitoring and Wearable Devices in Healthcare

Springer

Editors
Philip Eappen
Healthcare Management
Cape Breton University
Sydney, NS, Canada

Dimitrios Zikos
Texas Tech University Health Sciences
Center
Lubbock, TX, USA

Narasimha Rao Vajjhala
University of New York Tirana
Tirana, Albania

Karen Parker Davidson
Health Administration Division
School of Health Sciences
The Herbert H. & Grace A. Dow College
of Health Professions
Central Michigan University
Mount Pleasant, MD, USA

ISSN 3004-958X ISSN 3004-9598 (electronic)
Information Systems Engineering and Management
ISBN 978-3-031-98896-7 ISBN 978-3-031-98897-4 (eBook)
https://doi.org/10.1007/978-3-031-98897-4

This Springer imprint is published by the registered company Springer Nature Switzerland AG
The registered company address is: Gewerbestrasse 11, 6330 Cham, Switzerland

To my beloved father, Eappen—whose unwavering wisdom, strength, and compassion continue to illuminate my path. Though no longer with us, his values live on in every page of this work. This book is a tribute to his enduring spirit and to all those around the world who dedicate their lives to advancing care for patients and building healthier, more equitable communities.

Dr. Philip Eappen

Foreword

A Revolution at the Edge of the Body

In the early 21st century, a quiet revolution began in healthcare—one not driven by pharmaceutical breakthroughs or hospital expansions, but by devices no larger than a wristwatch or a patch. Today, wearable health devices—smartwatches, fitness bands, biosensors, and even smart fabrics—are reshaping how we understand, manage, and deliver care. Powered by the Internet of Things (IoT), these technologies have moved health monitoring from hospitals and clinics into homes, workplaces, and everyday life, enabling a level of continuous, personalized care that was once unimaginable.

This innovative book, *Remote Monitoring and Wearable Devices in Healthcare* edited by Dr. Eappen, arrives at a pivotal moment in this transformation. It offers a timely and authoritative exploration of the dynamic intersection between wearable technologies and the IoT. More than just a survey of gadgets, the book provides a critical framework to understand how these tools are reshaping clinical decision-making, public health surveillance, chronic disease management, and preventive medicine. As the healthcare sector increasingly embraces artificial intelligence, the insights offered in this volume will be essential for navigating the ethical, regulatory, and practical challenges of integrating real-time health data into everyday clinical practice.

What makes this book especially important is its ability to bridge disciplines. Drawing from technology, medicine, health policy, and ethics, Dr. Eappen presents a comprehensive analysis that speaks not only to healthcare professionals and technologists but also to policymakers, educators, and global health leaders. The book does not shy away from tough questions: How do we protect patient privacy in a world of constant data exchange? What are the risks of algorithmic bias in predictive tools? Can our health systems absorb and act upon decentralized data flows? And perhaps most importantly: Will wearable technologies narrow or widen health equity gaps?

I remain optimistic. Just as the mobile phone revolution brought communication and services to even the most underserved corners of the world, I believe these emerging health technologies can ultimately benefit all. In that spirit, this book is not only a guide for understanding today's wearable health revolution—it is a roadmap for ensuring that revolution is inclusive, ethical, and sustainable.

By illuminating both the promise and the perils of a new healthcare paradigm—one in which care is not only delivered to patients but also carried by or embedded within them—*Remote Monitoring and Wearable Devices in Healthcare* invites us to reimagine the very boundaries of care: from episodic to continuous, from reactive to predictive, from institutional to individual.

Welcome to the future of health—worn, wired, and always on.

Dr. William E. Bertrand
Professor
Tulane University
New Orleans, USA

Preface

The intersection of wearable technologies, remote monitoring systems, and intelligent health analytics marks one of the most transformative frontiers in modern healthcare. From managing chronic illness to enhancing provider well-being, from real-time diagnostics to empowering patient autonomy, these innovations are fundamentally reshaping care delivery in every region of the world.

Remote Monitoring and Wearable Devices in Healthcare brings together a diverse, global group of experts to examine these technologies from technical, clinical, ethical, and policy perspectives. This volume reflects the collaborative spirit of healthcare innovation spanning disciplines such as nursing, engineering, informatics, public health, and health administration and offers readers a rich exploration of the ways in which connected care is advancing.

The book's chapters delve into pressing topics: the role of the wearable Internet of Things (WIoT), machine learning's role in wearable data analysis, workplace integration of wearables, co-design in digital health development, and the profound implications of remote monitoring for chronic disease management, surgical assistance, and global health equity. These contributions reflect cutting-edge research from institutions across North America, Europe, Africa, and Asia.

Importantly, this book does not merely highlight technological breakthroughs—it interrogates them. The integration of wearable devices into healthcare systems demands rigorous attention to privacy, accessibility, health workforce readiness, and patient engagement. Our contributors confront these challenges head-on, offering practical frameworks, empirical evidence, and future directions.

We are deeply grateful to our chapter authors and reviewers who made this publication possible and to Springer for supporting this work. We also thank Prof. William E. Bertrand of Tulane University for his Foreword, which aptly situates this book within the broader movement toward digitally enabled, patient-centered healthcare.

As editors, we hope this book serves not only as a reference, but also as a catalyst for research, for policy change, and for the continued pursuit of equitable and intelligent healthcare for all.

Sydney, Canada — Dr. Philip Eappen
Tirana, Albania — Prof. Assoc. Narasimha Rao Vajjhala
Lubbock, USA — Prof. Assoc. Dimitrios Zikos
Mount Pleasant, USA — Dr. Karen Parker Davidson

Acknowledgments

We are honored to present *Remote Monitoring and Wearable Devices in Healthcare*, a collaborative effort shaped by the collective expertise, dedication, and vision of an international team of contributors. This book reflects not only technological progress but also a shared commitment to building a more responsive, inclusive, and intelligent healthcare system for the future.

I would like to express my deepest gratitude to all those who made this project possible. To my fellow editors—Assoc. Prof. Narasimha Rao Vajjhala, Assoc. Prof. Dimitrios Zikos, and Dr. Karen Parker Davidson—thank you for your collegiality, thoughtful insights, and unwavering support throughout this journey. Your leadership enriched this work and ensured its relevance across diverse healthcare contexts.

To the authors, your time, knowledge, and scholarship gave this book its substance. I am grateful to each of you for contributing chapters that explore not only technical innovation but also the ethical, social, and clinical dimensions of wearable technologies in healthcare.

I extend heartfelt appreciation to the institutions and organizations that supported this endeavor—your encouragement, resources, and belief in the project were essential.

Personally, I wish to thank my wife Figgi, son Evan, mother Baby, sister Annie, my in-laws, and my dear niece and nephew—for their enduring love, patience, and encouragement. I especially dedicate this book to the memory of my late father, Eappen, whose wisdom and values continue to guide and inspire me. This work is, in part, a tribute to him and to all those striving to improve care for patients and communities worldwide.

Dr. Philip Eappen

Contents

Editors and Contributors

About the Editors

Dr. Philip Eappen is an accomplished healthcare executive, academic, and researcher, currently serving as Assistant Professor of Healthcare Management at Cape Breton University. With over a decade of experience in hospital operations, healthcare education, and strategic leadership, he has consistently advanced innovation and excellence across diverse healthcare systems in Canada, the United States, and internationally.

Prior to joining Cape Breton University, Dr. Eappen held academic appointments at the University of Toronto, Southern Alberta Institute of Technology, and Fanshawe College. Internationally, he served as Director of Health Services and Chief Administrator of Healthcare Operations at the American University of Nigeria, where he led clinical transformation, emergency preparedness, and inclusive health initiatives, including programs supporting survivors of Boko Haram abductions.

Dr. Eappen holds a Doctorate in Healthcare Administration and a Post-Graduate Certificate in International Health from Central Michigan University, along with an MBA in Healthcare Management and a Bachelor of Nursing. He has taught and developed curricula across undergraduate and graduate levels, including healthcare informatics, organizational behavior, and health systems analytics, while leading the development of Cape Breton University's forthcoming Master of Healthcare Leadership and Innovation program.

An active contributor to national and global health dialogues, Dr. Eappen sits on the boards of the American College of Healthcare Executives (Canada Chapter), Myeloma Canada, Aplastic Anemia and Myelodysplasia Canada, and the Network of Rare Blood Disorder Organizations (NRBDO). He also serves on research and ethics committees, reflecting his commitment to governance, evidence-based practice, and health equity.

Dr. Eappen has published extensively on topics such as AI in healthcare, digital health transformation, and health workforce development. His recent editorial works include *Healthcare Informatics Innovation Post COVID-19 Pandemic* (Taylor &

Francis). He has presented at major international conferences including ICN 2025, ICIMTH, and INTED and is currently pursuing the Fellow of the American College of Healthcare Executives (FACHE) and Certified Health Executive (CHE) designations.

With a research agenda focused on digital innovation, rare diseases, and global workforce integration, Dr. Eappen continues to shape policy, practice, and pedagogy in healthcare management.

Dr. Narasimha Rao Vajjhala is a distinguished academic and researcher currently serving as Professor and Chair of the Department of Computer Science at the American University in Bulgaria (AUBG). He is also the Principal Investigator of the Democracy Project (DemPro) at the Center for Information, Democracy, and Citizenship (CIDC) at AUBG. With over two decades of experience in higher education, Dr. Vajjhala has held senior academic leadership positions, including Dean of the Faculty of Engineering and Architecture at the University of New York Tirana (UNYT), Albania, and Chair of the Computer Science and Software Engineering programs at the American University of Nigeria (AUN).

Dr. Vajjhala has taught a wide range of undergraduate and graduate courses in computer science, database systems, and programming across institutions in Europe and Africa. He is an active contributor to the global research community and serves as Editor-in-Chief of the *International Journal of Risk and Contingency Management (IJRCM)*. He is a Senior Member of both the Association for Computing Machinery (ACM) and the Institute of Electrical and Electronics Engineers (IEEE), and a member of the Project Management Institute (PMI).

In addition to his academic pursuits, Dr. Vajjhala has served as a technology consultant for various European firms and has been involved in numerous EU-funded research and innovation projects. He holds a Doctorate in Information Systems and Technology from the United States, an M.Sc. in Computer Science and Applications from India, and an MBA with a specialization in Information Systems from Switzerland. His interdisciplinary expertise spans computer science, information systems, and project management, with significant contributions to research, education, and industry practice.

Dr. Dimitrios Zikos is an Associate Professor in the Department of Healthcare Management and Leadership at Texas Tech University Health Sciences Center (TTUHSC). He is a core faculty member in the Master of Science in Healthcare Administration (MSHA) program, where he teaches graduate-level courses in health informatics, research methods, and statistics. Dr. Zikos brings extensive academic and leadership experience, having previously served as Director of the Health Administration Division and Interim Program Director of the Master of Health Administration at Central Michigan University. During his tenure there, he led the program's successful accreditation by the Commission on Accreditation of Healthcare Management Education (CAHME).

Dr. Zikos holds a Ph.D. in Health Informatics from the National and Kapodistrian University of Athens in Greece and a Bachelor's degree in Nursing. His professional background spans both Europe and the United States, with applied research and academic contributions focused on the intersection of health data, workforce management, and patient safety.

His research is centered on leveraging data analytics to improve clinical and administrative decision-making in healthcare. He specializes in the secondary analysis of medical claims data, comorbidity modeling, and the evaluation of inpatient outcomes such as mortality and length of stay. Dr. Zikos is particularly interested in the use of digital tools to enhance healthcare access for underserved populations, and the development of data-driven systems to support evidence-based care and decision-making. He is also active in the field of medical education technology, working on simulation-based tools that enhance the accessibility and usability of health information for both clinicians and patients.

A dedicated mentor, Dr. Zikos has chaired over twenty doctoral dissertations, supporting students who aim to apply informatics and analytics to solve complex, real-world healthcare challenges. He has authored numerous peer-reviewed articles, book chapters, and conference presentations in health informatics, healthcare management, and digital health systems.

Dr. Karen Parker Davidson, RN, DHA is a globally recognized Airway Clinical Nurse Specialist known for her pioneering work in nasal function and airway science. As one of only two experts worldwide advancing the clinical application of 4-phase rhinomanometry, she introduced the concept of nasal resistance as the "fifth vital sign"—an actionable, measurable biomarker in clinical care. Her research has fundamentally transformed how nasal airflow is assessed, enabling more precise diagnosis and treatment across sleep medicine, surgery, dentistry, and orthodontics.

Dr. Davidson is a prolific innovator and inventor, having developed a suite of proprietary clinical tools and methods, including the Davidson Maneuver for assessing the external nasal valve, the FACT Method™, the MAVERICK Nasal Exam™, and the DAFNE Score with Airway AI. These advancements are instrumental in diagnosing and managing nasal flow limitations and airway instability, particularly in patients with PTSD, bruxism, postural abnormalities, and dysfunctions induced by oral appliances or tongue positioning.

With over 30 years of experience in advanced practice nursing and the medical device industry, Dr. Davidson's clinical expertise spans critical care, oncology, infusion therapy, and military medicine. She served as a U.S. Air Force flight nurse and now leads as CEO of MedTec Research, Inc. and principal of FACT Healthcare Consulting Group. Her work has fostered high-level collaborations with leading global medical firms and academic institutions.

A committed educator and bestselling author, Dr. Davidson teaches at Liberty University and authored *Breathe Through Your Nose, Don't Pay Through It: The Impact the Healthcare Industry Has on Nasal Function and How We Breathe.* She is also the creator of the My Little Nose children's book series, promoting early education on nasal breathing. Her scholarly output includes numerous publications on airway diagnostics, craniofacial development, and the integration of remote monitoring technologies in healthcare.

Dr. Davidson's contributions have earned widespread recognition, including selection as one of Washington, D.C.'s Top Women Leaders (2022), the Outstanding Dissertation Award from Central Michigan University (2020), and a National Science Foundation award for innovation in emergency airway device development. In 2024, her team was commended by U.S. Speaker of the House Mike Johnson for their contributions to airway health. She was most recently honored as Professor of the Semester at Liberty University's School of Health Sciences for her leadership in healthcare policy.

Dr. Davidson holds a Doctor of Health Administration, three master's degrees, a Bachelor of Nursing and is currently pursuing a Ph.D. in Business Administration. She resides in Annapolis, Maryland.

Contributors

Hikmat Singh Brar Cape Breton University, Sydney, NS, Canada

Merritt Brockman Texas Tech University Health Sciences Center, Lubbock, TX, USA

Emma Brulin Unit of Occupational Medicine, Institute of Environmental Medicine, Karolinska Institute, Solna, Sweden

Rafiqul Chowdhury Department of Management Science, Shannon School of Business, Cape Breton University, Sydney, NS, Canada

Karen Parker Davidson Department of Public Health, School of Health Sciences, Liberty University, Lynchburg, VA, USA;
Division of Health Administration, School of Health Sciences, Central Michigan University, Mt. Pleasant, MI, USA;
MedTech Research, Gambrills, MD, USA

Philip Eappen School of Business, Cape Breton University, Sydney, NS, Canada

Pauline Frank Patient Co-Next, Zürich, Switzerland

Virginia Gunn Cape Breton University, Sydney, NS, Canada;
Unit of Occupational Medicine, Institute of Environmental Medicine, Karolinska Institute, Solna, Sweden

Hemant Gupta ICFAI Foundation for Higher Education, Bangalore, India

Kathleen Hodgkins Central Michigan University, Mount Pleasant, MI, USA

Md. Anisul Islam Department of Management Science, Shannon School of Business, Cape Breton University, Sydney, NS, Canada

Monirul Islam Associate Professor, Department of Management, Sister Nivedita University, Kolkata, India

Sai Sathish Kethu NeuraFlash, Burlington, USA

Gana Manjusha Kondepudi Department of Pharmacognosy, Vignan Institute of Pharmaceutical Technology, Visakhapatnam, Andhra Pradesh, India

Carles Muntaner Lawrences F. Bloomberg Faculty of Nursing and Dalla Lana School of Public Health, University of Toronto, Toronto, Canada

Radhika Ramachandran Nair Department of Management Science, Shannon School of Business, Cape Breton University, Sydney, NS, Canada

Swapna Narla Tek Yantra Inc, Folsom, CA, USA

Durai Rajesh Natarajan Estrada Consulting Inc, Sacramento, CA, USA

Bhavesh Kumar J. Parmar School of Management Studies, National Forensic Sciences University, Gandhinagar, India

Sreekar Peddi Tek Leaders, Plano, TX, USA

Vijay Srinivas Pothula Stira Pharma LLC, Fairfield, New Jersey, USA

Gadha Jyothi Prasad Department of Management Science, Shannon School of Business, Cape Breton University, Sydney, NS, Canada

N. Purandhar Department of CSE (Artificial Intelligence), School of Computers, Madanapalle Institute of Technology and Science, Madanapalle, Andhra Pradesh, India

Charu Saxena Faculty of Pharmacy, Department of Pharmaceutics, Swami Vivekanand Subharti University Meerut, Meerut, India

Wasswa Shafik Dig Connectivity Research Laboratory (DCRLab), Kampala, Uganda;
School of Digital Science, Universiti Brunei Darussalam, Gadong, Brunei

Kalgi Shah School of Management Studies, National Forensic Sciences University, Gandhinagar, India

Kanikella Sivaji Department of Pharmacology, AKRG College of Pharmacy, Nalljerla, Andhra Pradesh, India

Dharma Teja Valivarthi Tek Leaders, Plano, TX, USA

Vikrant Verma Faculty of Pharmacy, Department of Pharmaceutical Chemistry, Swami Vivekanand Subharti University Meerut, Meerut, India

William R. Young University of Exeter, Exeter, UK

Dimitrios Zikos Texas Tech University Health Sciences Center, Lubbock, TX, USA

The Economic and Health Policy Impact of Wearable Devices in Healthcare

Karen Parker Davidson

Abstract Wearable devices are transforming healthcare by providing real-time health monitoring, improving patient engagement, and offering data-driven insights for clinicians. This chapter explores the economic and health policy implications of wearable technology in modern healthcare systems. From reducing hospital readmissions to enabling personalized care, wearables present significant opportunities for cost savings and enhanced health outcomes. The chapter examines key economic benefits, including optimized resource allocation, improved chronic disease management, and reduced healthcare disparities through remote care access. Furthermore, it addresses the challenges posed by wearables, such as data privacy concerns, integration with existing healthcare systems, and regulatory hurdles. Policy considerations focus on reimbursement models, equitable access, and standardizing data interoperability to maximize the potential of wearable devices. Case studies are presented to illustrate the impact of wearables on population health and healthcare delivery. By analyzing these economic and policy dimensions, the chapter highlights how wearable technology is reshaping healthcare, fostering innovation, and enabling a more patient-centered approach to care. This exploration underscores the need for collaborative efforts among stakeholders to harness the full potential of wearable devices for sustainable and equitable healthcare transformation.

Keywords Healthcare · Policy · Economies · Wearable devices · Remote monitoring · Reimbursement

K. P. Davidson (✉)
Department of Public Health, School of Health Sciences, Liberty University, Lynchburg, VA, USA
e-mail: kpdavidson@liberty.edu

Division of Health Administration, School of Health Sciences, Central Michigan University, Mt. Pleasant, MI, USA

MedTech Research, Gambrills, MD, USA

P. Eappen et al. (eds.), *Remote Monitoring and Wearable Devices in Healthcare*,
Information Systems Engineering and Management 63,
https://doi.org/10.1007/978-3-031-98897-4_1

1 Introduction

The rapid advancement of wearable technology has brought a paradigm shift in healthcare, offering unprecedented opportunities to improve health outcomes, enhance patient engagement, and reduce healthcare costs. These devices, which range from fitness trackers and smartwatches to sophisticated medical-grade sensors, have become integral tools in monitoring and managing a wide range of health conditions. By providing real-time data on metrics such as heart rate, activity levels, sleep patterns, and even blood oxygen saturation, wearables empower individuals to take a proactive role in their health while enabling clinicians to make more informed decisions.

This chapter examines the economic and health policy implications of wearable devices, focusing on their potential to transform healthcare delivery. The adoption of wearable technology has demonstrated significant benefits, including the ability to predict and prevent medical emergencies, manage chronic diseases, and support population health initiatives. However, the integration of these devices into healthcare systems is not without challenges. Issues such as reimbursement models, equitable access, data privacy, and interoperability present barriers that must be addressed to unlock their full potential.

In addition to exploring these challenges, this chapter highlights policy considerations that can promote the equitable and effective adoption of wearable technology. By analyzing real-world case studies and examining the interplay between technology, economics, and policy, the discussion sheds light on how wearable devices can be leveraged to create a more patient-centered and efficient healthcare system. Through these insights, the chapter aims to provide a comprehensive understanding of how wearable devices are shaping the future of healthcare while addressing the critical economic and policy frameworks needed to support their integration.

1.1 Economic Impact of Wearable Devices in Healthcare

The adoption of wearable devices in healthcare has transformed how medical care is delivered, monitored, and evaluated [1]. These devices, ranging from fitness trackers to advanced medical-grade monitors for sleep, oxygen saturation, heart rate variability (HRV), cognitive behavioral therapy (CBT), monitoring of chronic disease management, and breathing, to name a few, enable real-time data collection and remote monitoring of patients [2].

This chapter explores the economic implications of wearable healthcare technologies through a cost–benefit analysis, their impact on healthcare efficacy, and the policy considerations that accompany their widespread use. Furthermore, it delves into the total cost savings generated by these devices compared to traditional in-office visits and their role in enhancing access to care within the global market. The global wearable healthcare technology market has experienced substantial growth,

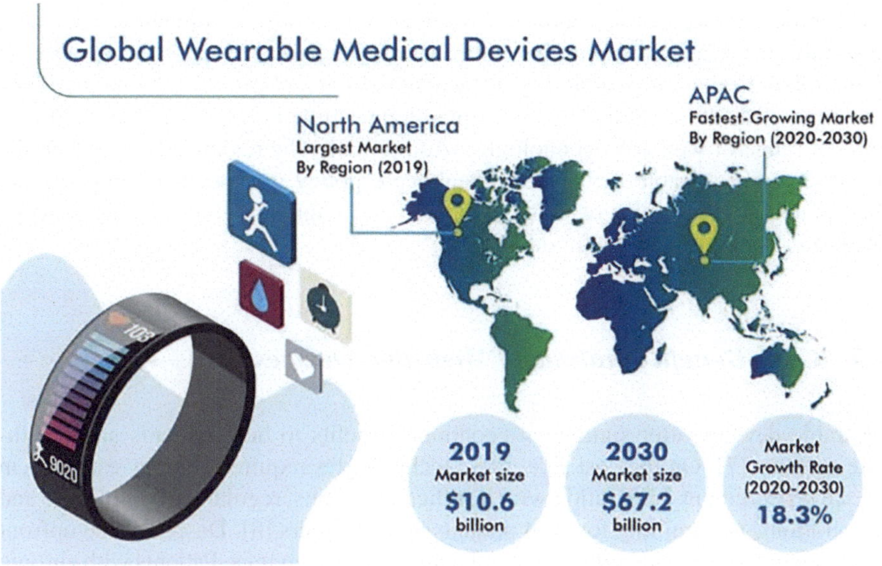

Fig. 1 Global wearable device market

with projections estimating it to reach $46.6 billion by 2025, reflecting a compound annual growth rate of 23.3%7, and $67.2 billion by 2030 as seen in Fig. 1 [3].

1.2 Cost–Benefit Analysis of Wearable Devices

Over the past five years, wearable technology has emerged as a significant player in healthcare economics, particularly in the realm of symptom monitoring. This growth is driven by the increasing focus on preventive healthcare and the rise of chronic diseases worldwide. The integration of artificial intelligence and machine learning into wearables has enhanced their capabilities, enabling more accurate and personalized health insights, indicating substantial cost savings for healthcare systems. However, challenges such as data privacy concerns and the need for robust clinical validation remain hurdles to widespread adoption and integration into standard healthcare practices.

Many papers have examined and analyzed the concept of wearables and the economic impact as recent as 2024 [4]. The findings highlighted that wearable technologies have the potential to significantly enhance healthcare outcomes, particularly by improving quality-adjusted life years across various patient demographics and medical conditions [5]. The analysis underscored several factors that influence the cost-effectiveness of these technologies, including the type of device, the health conditions they target, economic perspectives, and regional cost structures.

The authors stressed the importance of conducting more comprehensive research to solidify the evidence supporting the economic benefits of wearables, in addition to the reliability and dependability of the technology and product claims [6]. They recommended future studies that examined longer time horizons and encompass a broader range of wearable technologies. Additionally, the review advocated for the integration of wearable devices into healthcare policy and practice, emphasizing their potential to optimize resource allocation and improve patient outcomes while ensuring strong protections for data privacy.

1.3 Cost–Benefit Analysis of Wearable Devices

Wearable devices offer significant economic benefits to both patients and healthcare systems [7]. On the cost side, these technologies require initial investments in device procurement, integration with healthcare systems, regulatory compliance and registrations, and training for both patients and providers [8]. Despite these upfront costs, wearables often result in substantial downstream savings. Patients with chronic diseases, such as diabetes or hypertension, can use wearable devices to monitor their conditions in real-time, reducing the need for frequent clinic visits and emergency interventions. For example, continuous glucose monitors (CGMs) allow diabetics to self-manage their blood sugar levels, minimizing complications and hospitalizations.

On the benefit side, wearable devices improve patient outcomes by facilitating early detection and prevention of medical issues [9]. Remote monitoring technologies, such as wearable ECGs or blood pressure monitors, alert healthcare providers to potential problems before they escalate. These early interventions not only save lives but also reduce the financial burden of treating advanced-stage diseases. Employers and insurers have also embraced wearables to incentivize healthy behavior among employees and beneficiaries, creating additional economic benefits by reducing absenteeism and improving workplace productivity.

Wearable devices have shown potential to reduce healthcare costs significantly, with estimates suggesting they could lower hospital costs by up to 16% over five years. Wearable devices also have implications for preventive care, enabling individuals to proactively address their health [10, 11]. For instance, fitness trackers and wearable devices that monitor activity levels, sleep quality, and heart rate encourage users to adopt healthier lifestyles [12]. This shift towards preventive care can reduce the prevalence of lifestyle-related diseases such as obesity, cardiovascular conditions, and type 2 diabetes [13]. In turn, the reduced disease burden leads to decreased healthcare spending at both individual and systemic levels [14].

Moreover, wearable devices streamline data collection, facilitating the integration of patient information into electronic health records (EHRs) or electronic medical records (EMR) [15]. This integration enhances the efficiency of healthcare delivery by providing providers with comprehensive patient data. It also minimizes redundancies and errors associated with manual data entry, contributing to cost savings and improved patient safety. Wearables enable providers to monitor health trends over time, identify patterns, and tailor interventions to individual needs, further increasing their value proposition.

In a comprehensive review of wearable devices, one group analyzed the economic outcomes among If we examined the various forms of monitoring, the probability of cost savings is 66–100% [4]. The summarized analysis highlights the cost-effectiveness of various wearable health devices across different applications. Studies have shown that wearable devices, such as the RespiraSense respiratory monitor and NightWatch seizure detection device, often deliver better clinical outcomes at lower costs compared to traditional methods or no intervention [16]. For example, RespiraSense proved dominant by reducing costs (£221.4 savings per patient) and improving QALYs (6.926 vs. 6.917), while NightWatch was cost-effective at €846 per patient with significant caregiver benefits [4].

Pedometers and wearable systems for managing hospital-acquired pressure injuries (HAPIs) demonstrated notable cost savings and health benefits. Pedometers provided life-year and QALY gains at €11,100 per QALY, while HAPI prevention systems reduced costs by $6621 per patient annually and achieved a 77% decrease in injury incidence [17]. Conversely, interventions like sensor-augmented insulin therapy and wearable cardioverter-defibrillators (WCDs) exhibited higher costs but still met thresholds for cost-effectiveness in specific scenarios, such as home monitoring versus inpatient care.

Key findings emphasize that wearable technologies can improve healthcare outcomes and quality-adjusted life years (QALYs) while being economically favorable under established willingness-to-pay thresholds. However, the cost-effectiveness of some devices depends on implementation settings, long-term benefits, and device pricing. These insights underscore the importance of integrating cost-effectiveness assessments into the development and deployment of wearable health devices.

1.4 Listen to Your Heart: It Could Save You Money

One of the most common and oldest forms of monitoring is Holter monitoring as seen in Fig. 2 [18]. Holter monitors as shown in Exhibit 10.2, and in-lab cardiac tests represent two approaches to diagnosing and managing cardiac conditions, each with distinct economic implications.

A Holter monitor, typically used for ambulatory ECG monitoring over 24–48 h, allows patients to continue their daily activities while collecting data on heart rhythm and electrical activity. The cost of a Holter monitor is lower than that of in-lab tests due to the elimination of facility overheads and the reduced need for specialized staff during the monitoring period. The average cost for a Holter monitor study ranges between $150 and $600, depending on the healthcare provider and geographic location [19]. Since it provides continuous data over an extended period, it offers a comprehensive picture of intermittent arrhythmias without requiring prolonged clinical visits.

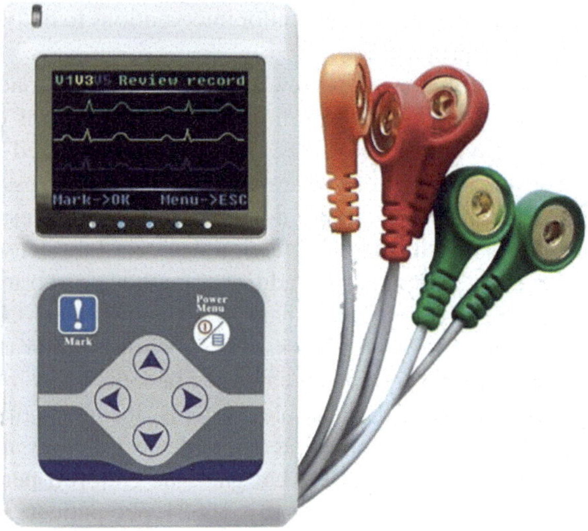

Fig. 2 Holter monitor

In contrast, in-lab cardiac tests, such as treadmill stress tests, echocardiograms, or electrophysiology studies, are conducted in a controlled clinical environment. These tests require specialized equipment, trained technicians, and sometimes sedation or contrast agents. The costs for such tests can range from \$1000 to \$5000 or more, depending on the procedure [20]. While these tests offer immediate, real-time data under controlled conditions, they may fail to capture transient cardiac events that occur outside the clinical setting.

From a cost–benefit perspective, Holter monitors are more economical for initial diagnostics of episodic symptoms like palpitations or dizziness [21]. They also reduce the burden on healthcare systems by minimizing the need for multiple clinic visits and lowering resource utilization. However, in-lab tests remain indispensable for comprehensive cardiac assessments, such as evaluating structural abnormalities or precise measurements of cardiac function, which a Holter monitor cannot provide.

Moreover, the use of Holter monitors aligns with trends in remote monitoring and wearable technology, enabling healthcare providers to detect and manage cardiac conditions earlier. This approach reduces hospital admissions and emergency interventions, contributing to long-term cost savings. By comparison, in-lab tests are often used after an initial diagnosis to provide deeper insights or guide interventions, making them critical but less frequent.

Both methods have their place in a comprehensive cardiac care strategy. However, the economic efficiency of Holter monitors for widespread use in population health management highlights their value as a cost-effective tool for early detection and monitoring.

1.5 History of Sleep Apps for Remote Monitoring: Is There an Economic Impact?

In conjunction with cardiac monitoring in cardiology, there is monitoring of sleep and airway function. Figure 3 is an example of one of many sleep apps that have evolved significantly over the past two decades, reflecting advances in smartphone technology, wearable devices, and medical-grade monitoring tools. Early sleep apps, emerging around the late 2000s, were primarily focused on sleep tracking through motion detection using smartphone accelerometers. These apps, such as Sleep Cycle and Pillow, analyzed sleep patterns by tracking movements during the night, offering users basic insights into their sleep duration and quality. While informative, their accuracy was limited compared to clinical methods.

With the rise of wearable devices in the 2010s, such as Fitbit, Garmin, and Apple Watch, sleep tracking became more sophisticated. These devices combined accelerometers, heart rate monitors, and, eventually, SpO2 sensors to assess sleep stages and identify disruptions. The integration of these wearables with companion apps enhanced their functionality, allowing for more precise data collection and real-time feedback. Some wearables, like the Oura Ring and WHOOP strap (Exhibit 10.4), focused specifically on recovery and sleep quality, catering to users seeking detailed insights for health optimization.

By the early 2020s, sleep apps began incorporating medical-grade functionalities [22]. Apps like SleepScore and Sleepio started leveraging artificial intelligence (AI) to analyze breathing patterns, snoring, and environmental factors affecting sleep. These apps also offered cognitive behavioral therapy for insomnia (CBT-I), which

Fig. 3 Sleep app

has been recognized as a gold-standard treatment for chronic sleep problems. The advent of contactless sleep tracking using radar-based systems (e.g., Google Nest Hub) further expanded the non-invasive monitoring options available to consumers.

More recently, sleep apps have integrated with telemedicine platforms, enabling remote consultations with sleep specialists [23]. They are also being used in clinical trials and research to evaluate sleep disorders like obstructive sleep apnea (OSA). The development of algorithms capable of detecting conditions such as OSA has pushed these apps into the realm of remote healthcare monitoring, providing a cost-effective alternative to traditional in-lab testing [24].

1.6 Cost-Effectiveness of Sleep Apps Versus In-Lab Testing

In-lab sleep testing, or polysomnography (PSG), is considered the gold standard for diagnosing sleep disorders [25]. It involves an overnight stay in a sleep clinic, where multiple parameters—including brain activity, eye movement, heart rate, breathing, and oxygen saturation—are recorded. The cost of a single PSG can range from $1000 to $5000, depending on the complexity of the test and the healthcare facility [26]. In-lab testing is comprehensive but costly and resource-intensive, requiring specialized equipment, trained technicians, and dedicated sleep facilities.

In contrast, home sleep testing, sleep apps, and wearable devices are significantly more cost-effective. The initial investment for a wearable device ranges from $100 to $400, with some apps requiring additional subscription fees of $10 to $50 per month. Over time, the cumulative cost of using home sleep testing remains much lower than that of in-lab testing, especially for individuals needing ongoing monitoring. Figure 4 is an example of a home sleep test wearable device.

Advantages of Sleep Apps for Cost-Effectiveness:

Fig. 4 Home sleep test, WatchPat. Source Zoll Medical, Inc

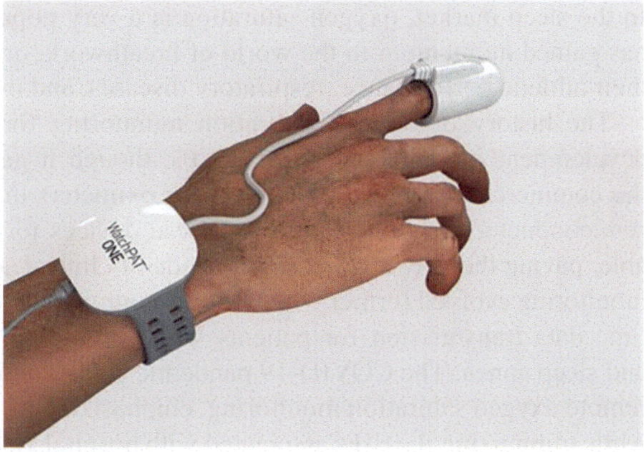

Fig. 5 Home sleep test

1. Accessibility and Scalability: Sleep apps and wearables are widely accessible, enabling large-scale monitoring at a fraction of the cost of in-lab testing. This is especially valuable in regions with limited access to sleep clinics.
2. Convenience: Users can monitor their sleep at home without the discomfort or logistical challenges of traveling to a clinic. This encourages long-term adherence to monitoring and self-management.
3. Preventive Approach: Sleep apps empower users to identify and address sleep problems early, potentially reducing the need for expensive clinical interventions.
4. Telemedicine Integration: The ability to share app data with healthcare providers facilitates remote consultations, further reducing costs associated with in-person visits (Fig. 5).

Limitations and Complementary Roles

While sleep apps and wearables are cost-effective, their accuracy in diagnosing complex sleep disorders like OSA or periodic limb movement disorder remains inferior to PSG. However, they are effective for screening, monitoring treatment adherence (e.g., CPAP compliance), and tracking general sleep trends. In cases where sleep apps identify potential issues, a PSG may still be warranted, making these tools complementary rather than replacements for in-lab testing.

1.7 Oxygen Saturation, Sensing, and Monitoring

Sleep apps have revolutionized remote sleep monitoring, offering a convenient, accessible, and cost-effective alternative to in-lab testing. While they cannot entirely replace the diagnostic capabilities of polysomnography, their role in preventive care and ongoing monitoring has made them invaluable in modern healthcare. By reducing the reliance on resource-intensive sleep clinics, these technologies contribute to significant healthcare savings while improving access to sleep health management.

In the sleep market, oxygen saturation is a very popular wearable technology that has gained momentum in the world of breathwork, or for those looking to improve their athletic performance, respiratory diseases, and overall well-being.

The history of oxygen saturation monitoring for remote use began with the development of oximetry in the 1940s, though it gained widespread utility with the commercialization of portable pulse oximeters in the 1980s [27]. Advances in semiconductor technology allowed these devices to become compact and affordable, paving the way for their use outside of clinical settings. By the 2000s, remote monitoring evolved further with wireless communication capabilities, enabling real-time data transmission for patients with chronic conditions like COPD, asthma, and sleep apnea. The COVID-19 pandemic significantly accelerated the adoption of remote oxygen saturation monitoring, emphasizing its role in tracking patient health while minimizing the risks associated with hospital visits [27]. In recent years, wearable devices such as smartwatches have incorporated oxygen saturation monitoring, increasing accessibility to preventive care, though their accuracy often falls short of clinical-grade devices.

In comparing remote monitoring to in-lab testing, remote systems prove to be more cost-effective for chronic disease management and large-scale screening. Remote monitoring reduces healthcare costs by minimizing hospitalizations and travel expenses, while also allowing for continuous data collection that facilitates early detection of hypoxemia. Devices for remote monitoring are relatively affordable, ranging from $20 to $300, with subscription services adding $10 to $50 per month for more advanced features. However, these systems rely heavily on patient compliance and proper usage, and their accuracy can vary, especially in consumer-grade devices.

In contrast, in-lab testing, such as polysomnography, provides unparalleled accuracy and diagnostic capability, as it is conducted in controlled environments using calibrated equipment. This method is ideal for complex cases and simultaneous monitoring of multiple health parameters. However, in-lab testing incurs significantly higher costs, ranging from $1000 to $3000 per session, and poses logistical challenges for patients, such as travel and overnight stays. Additionally, limited availability in rural or underserved areas further restricts its accessibility [28].

From a healthcare system perspective, remote monitoring is a more cost-effective solution for managing chronic conditions and triaging patients who may require in-lab testing. Studies have shown that it can reduce hospital readmissions by 20–30%, offering significant savings over time. In-lab testing, while more expensive, remains the gold standard for accurate diagnoses [29]. Together, these approaches offer complementary benefits, with remote monitoring providing scalable and affordable care while in-lab testing addresses more complex diagnostic needs.

A new wearable technology called SONU from Sound Health is making its way through the airway and nasal congestion market. This is a combination of artificial intelligence (AI) and a phone app.

1.8 Healthcare Efficacy and Reduction in Hospital Admissions

The efficacy of wearable devices in improving healthcare outcomes cannot be overstated. By providing continuous, non-invasive monitoring, wearables enable healthcare providers to personalize treatment plans and make data-driven decisions. This approach leads to better disease management and adherence to treatment protocols. Remote monitoring solutions, particularly for chronic diseases, have been shown to significantly reduce hospital admissions and readmissions. For instance, heart failure patients monitored remotely through wearable devices experience fewer hospitalizations compared to those relying solely on in-office check-ups. Similarly, patients recovering from surgeries benefit from wearable technologies that track vital signs and flag potential complications, reducing the likelihood of hospital readmission.

Additionally, wearable devices empower patients to take an active role in their health management. By providing actionable insights through user-friendly interfaces, these devices enhance patient engagement and compliance with prescribed regimens. Increased engagement has been linked to better clinical outcomes, translating into economic savings for healthcare systems. The reduction in hospital admissions not only alleviates the financial strain on healthcare facilities but also allows providers to allocate resources more effectively, improving overall system efficiency.

Remote monitoring is particularly impactful in managing chronic diseases, which are among the leading drivers of healthcare costs worldwide. Wearable devices designed for chronic disease management provide real-time data on parameters such as blood glucose levels, blood pressure, and oxygen saturation. These data enable healthcare providers to adjust treatment plans promptly, preventing disease progression and complications. For example, wearable cardiac monitors can detect arrhythmias early, allowing for timely interventions that prevent strokes or heart attacks. The economic benefits of such interventions are significant, as they avoid costly emergency care and long-term rehabilitation.

The integration of wearable devices into telemedicine platforms further enhances their efficacy. Patients in remote or underserved areas can access specialized care without traveling long distances. This not only reduces transportation costs but also ensures patients receive timely interventions. Telemedicine combined with wearable technology has been particularly beneficial during public health crises, such as the COVID-19 pandemic, when minimizing in-person interactions was critical. The ability to monitor patients remotely while maintaining high-quality care underscores the value of wearables in ensuring healthcare continuity.

Wearable home monitoring devices have emerged as transformative tools in healthcare, offering significant potential to reduce hospital admission rates and improve cost-effectiveness. These devices, which include smartwatches, fitness trackers, and medical-grade wearables, enable real-time monitoring of vital signs such as oxygen saturation, heart rate, respiratory rate, and even ECG. By providing continuous data outside clinical settings, wearable monitoring facilitates early detection of health issues, enabling timely interventions that can prevent the progression of conditions and reduce the need for hospitalizations.

Several studies have demonstrated the effectiveness of wearable devices in lowering hospital admission rates, particularly among patients with chronic conditions like heart failure, chronic obstructive pulmonary disease (COPD), and diabetes [29]. For instance, remote monitoring of patients with heart failure has been associated with a 25–40% reduction in hospital readmissions [30]. Similarly, COPD patients using wearable oxygen saturation monitors have experienced fewer exacerbations requiring hospitalization. These outcomes are primarily driven by wearable devices' ability to identify subtle changes in health metrics that may signal an impending crisis, allowing for early outpatient treatment.

From a cost-effectiveness perspective, wearable home monitoring is particularly advantageous. The cost of these devices ranges from $50 for consumer-grade models to $300–$500 for medical-grade wearables, with additional subscription services costing around $10–$50 per month [31]. Compared to the average cost of a hospital admission, which can range from $10,000 to $30,000 depending on the condition, wearable monitoring provides a substantial economic benefit [32]. Studies estimate that remote monitoring programs can reduce healthcare costs by 15–30% by avoiding unnecessary hospital stays, optimizing resource allocation, and improving overall patient outcomes [33].

Moreover, wearable monitoring enhances patient engagement in their health management, leading to better adherence to treatment plans and lifestyle modifications. This proactive approach further reduces the likelihood of hospitalizations. For healthcare systems, the scalability of wearable technology makes it an attractive solution for managing large patient populations, particularly in resource-constrained settings.

However, the cost-effectiveness of wearable monitoring depends on several factors, including device accuracy, patient compliance, and the integration of data into clinical workflows. While consumer-grade devices are affordable, they may lack the precision required for clinical decision-making. On the other hand, medical-grade wearables, though more expensive, offer higher reliability and better outcomes when coupled with robust telemedicine support. In conclusion, wearable home monitoring significantly lowers hospital admission rates and proves to be a cost-effective strategy for managing chronic conditions, improving patient outcomes, and reducing the financial burden on healthcare systems.

1.9 Policy Implications

The integration of wearable devices into mainstream healthcare raises several policy considerations. Policymakers must address issues related to data privacy, device regulation, and reimbursement models to maximize the potential benefits of these technologies. Ensuring the security of patient data collected by wearable devices is paramount. Robust regulations and encryption standards are necessary to protect sensitive health information from breaches and misuse.

From a regulatory standpoint, establishing clear guidelines for the approval and use of wearable medical devices is essential. The rapid pace of technological innovation often outstrips the ability of regulatory bodies to evaluate these devices. Streamlined pathways for device approval, coupled with rigorous safety and efficacy standards, will encourage innovation while ensuring patient safety. Reimbursement policies also play a crucial role in the adoption of wearable technologies. Many healthcare systems and insurance providers are still navigating how to adequately compensate providers for remote monitoring services. Establishing fair reimbursement models that reflect the cost savings and improved outcomes associated with wearable devices will incentivize their broader adoption.

Policymakers must also consider the equity implications of wearable technology. While these devices hold great promise for improving healthcare access and outcomes, their benefits may not be evenly distributed. Vulnerable populations, including low-income individuals and those in rural areas, may face barriers to accessing wearable devices due to cost or lack of technological infrastructure. Addressing these disparities requires targeted policies, such as subsidies for low-income patients or investments in broadband infrastructure to support telehealth services.

The development of standards for data interoperability is another critical policy area. Wearable devices from different manufacturers often use proprietary systems that are not easily integrated with each other or with existing EHRs. Establishing common standards for data sharing and interoperability will enhance the utility of wearable devices by enabling seamless communication between devices and healthcare systems. This, in turn, will facilitate more comprehensive and coordinated care.

Total Cost Savings and Access to Care

One of the most compelling economic arguments for wearable devices is their potential to reduce overall healthcare costs. By shifting care from expensive, resource-intensive in-office visits to cost-effective remote monitoring solutions, wearable devices generate substantial savings. Studies have shown that patients using remote monitoring technologies experience up to 30% fewer hospitalizations and emergency room visits, leading to significant reductions in healthcare expenditures. These savings extend beyond direct medical costs, encompassing reduced transportation expenses, lost productivity, and caregiver burden.

Moreover, wearable devices enhance access to care, particularly for underserved populations and those in rural areas. Remote monitoring eliminates geographical barriers, allowing patients to receive high-quality care regardless of location. This increased accessibility not only improves health equity but also reduces the economic disparities associated with healthcare access. Telehealth platforms integrated with wearable devices further amplify these benefits by enabling seamless communication between patients and providers.

Wearable devices also contribute to cost savings by reducing the need for diagnostic tests and imaging studies. Continuous monitoring provides a wealth of data that can be used to assess patient health without requiring additional tests. For example, wearable devices that track heart rate variability and activity levels can provide valuable insights into cardiovascular health, reducing the need for routine stress tests or Holter monitor studies. These savings are particularly significant for healthcare systems with limited resources, as they allow providers to focus on high-priority cases.

In addition to direct cost savings, wearable devices have broader economic benefits. By promoting better health outcomes and reducing the burden of chronic diseases, these technologies contribute to a healthier workforce. This, in turn, enhances productivity and economic growth. Employers who invest in wearable devices for their employees often report lower healthcare costs and higher employee satisfaction, further underscoring the economic value of these technologies.

In conclusion, wearable devices represent a transformative force in healthcare economics. Through cost-effective remote monitoring solutions, improved healthcare efficacy, and reduced hospital admissions, these technologies deliver substantial economic and clinical benefits. Policymakers must address challenges related to data privacy, regulation, and reimbursement to fully realize the potential of wearables. By doing so, they can unlock significant cost savings while enhancing access to care, ultimately advancing the goals of a more efficient, equitable, and patient-centered healthcare system.

Health Policy and Reimbursement for Wearable Technologies

The integration of wearable technologies into healthcare has revolutionized how medical conditions are managed, monitored, and prevented. Devices such as fitness trackers, smartwatches, and medical-grade wearables have enabled real-time monitoring of vital signs, improved patient engagement, and supported early intervention strategies. However, their widespread adoption hinges on robust health policies and reimbursement frameworks that incentivize both patients and providers while ensuring equitable access. This discussion explores the interplay between wearable technologies, health policy, and reimbursement models, examining their impact on healthcare delivery and potential policy advancements.

Challenges in Policy and Reimbursement

While the benefits of wearable technologies are well-documented, their adoption faces several barriers, including regulatory challenges, inconsistent reimbursement policies, and questions regarding data privacy and security, further affecting the economics. These challenges must be addressed to maximize the potential of wearables in transforming healthcare.

1. *Regulatory Frameworks*:

    - The classification of wearables as medical devices varies by jurisdiction, impacting the rigor of regulatory requirements.

- In the United States, the Food and Drug Administration (FDA) oversees medical-grade wearables, requiring evidence of safety and efficacy. Consumer-grade devices often fall outside FDA regulation, raising concerns about accuracy and reliability.
- Internationally, the European Union's Medical Device Regulation (MDR) and other regional frameworks govern wearable devices, creating a patchwork of standards that manufacturers must navigate.

2. *Reimbursement Policies*:
   - Reimbursement for wearable technologies is inconsistent, often limiting their adoption.
   - Public payers like Medicare and Medicaid in the U.S. have specific criteria for covering remote monitoring devices, often requiring prior authorization and evidence of medical necessity.
   - Private insurers vary in their coverage policies, with some offering reimbursement for devices that demonstrate cost savings or improved outcomes, while others exclude them altogether.
   - Internationally, reimbursement frameworks are even less developed, with many countries lacking formal policies for wearables.
3. *Data Privacy and Security*:
   - Wearable technologies generate vast amounts of personal health data, raising concerns about data breaches and misuse adding to cost of the device.
   - Policies such as the Health Insurance Portability and Accountability Act (HIPAA) in the U.S. provide a foundation for protecting patient data but may not fully address the complexities of wearable data.
   - Patients and providers often face challenges in ensuring compliance with privacy regulations while leveraging wearable data for care.

Health Policy Considerations for Wearable Technologies

To unlock the full potential of wearable technologies, policymakers must address the regulatory and reimbursement challenges while fostering innovation and equitable access as it relates to economics, access, and equity among all populations in various geographies. Key considerations include:

1. *Establishing Clear Regulatory Standards*:
   - Policymakers should develop unified standards for wearable technologies, ensuring safety and efficacy without stifling innovation.
   - International harmonization of regulations could streamline the approval process for manufacturers and expand access to wearables globally.
2. *Expanding Reimbursement Models*:
   - Value-based care models that tie reimbursement to patient outcomes can incentivize the adoption of wearable technologies.

- Public and private payers should consider covering devices that demonstrate cost savings, improved clinical outcomes, or enhanced patient engagement.
- Bundled payment models, where providers receive a single payment for comprehensive care episodes, could include wearable technologies as part of remote monitoring and chronic disease management programs.

3. *Promoting Data Interoperability and Security*:

   - Interoperability standards should enable seamless integration of wearable data into electronic health records (EHRs), enhancing care coordination. This could add to the cost of the EHR systems at the office or hospital system levels.
   - Robust cybersecurity measures and updated privacy policies are essential to protect patient data while allowing its use for personalized care and research.

4. *Addressing Equity and Accessibility*:

   - Subsidies or public funding programs could ensure that wearable technologies are accessible to underserved populations.
   - Policymakers should address digital literacy gaps and provide support for patients unfamiliar with wearable devices.

Reimbursement Models for Wearable Technologies

Reimbursement frameworks play a pivotal role in determining the adoption and utilization of wearable technologies. Several emerging models demonstrate how payers and providers can incentivize the use of wearables in clinical care.

1. *Fee-for-Service (FFS) Models*:

   - Under FFS models, providers bill for each service rendered, including remote monitoring enabled by wearables.
   - For example, Medicare in the U.S. offers reimbursement codes for Remote Patient Monitoring (RPM) services, covering the use of wearable devices for chronic disease management.

2. *Value-Based Care Models*:

   - These models tie reimbursement to patient outcomes, encouraging providers to adopt technologies that improve health metrics.
   - Wearables that reduce hospital readmissions or improve chronic disease management align well with value-based care goals, providing both clinical and economic benefits.

3. *Bundled Payments*:

   - In this model, providers receive a single payment for an entire episode of care, incentivizing the use of cost-effective solutions like wearables to manage patient outcomes.

- For instance, wearables used in post-operative monitoring could reduce complications and hospital readmissions, improving the cost-effectiveness of bundled payments.

4. *Insurance Incentives*:
   - Some private insurers offer premium discounts or incentives for members who use wearable devices to track and improve their health metrics.
   - These programs not only promote healthier behaviors but also reduce long-term healthcare costs for insurers.

1.10 Economic and Clinical Impact

The economic and clinical benefits of wearable technologies underscore the importance of supportive policies and reimbursement frameworks. Wearables have been shown to:

1. **Reduce Hospitalizations**:
   - Remote monitoring programs leveraging wearables have reduced hospital admissions for conditions like heart failure by 25–40%, generating significant cost savings.
2. **Improve Chronic Disease Management**:
   - Continuous monitoring of conditions such as diabetes and hypertension enables timely interventions, preventing complications and lowering healthcare costs.
3. **Enhance Preventive Care**:
   - By promoting healthier behaviors and detecting early warning signs of disease, wearables contribute to improved population health and reduced healthcare spending.
4. **Support Health Equity**:
   - When accessible to underserved populations, wearable technologies can bridge gaps in care, improving outcomes and reducing disparities.

1.11 Future Directions in Policy and Reimbursement

To fully integrate wearable technologies into healthcare systems, future policies should focus on:

1. **Innovative Payment Models**:

- Policymakers should explore innovative payment models that align financial incentives with the adoption of wearables, such as shared savings programs.

2. **Public–Private Partnerships**:
   - Collaboration between governments, insurers, and technology companies can drive innovation while ensuring equitable access and affordability.
3. **Education and Awareness**:
   - Educating patients and providers about the benefits and proper use of wearable technologies is essential to maximize their impact.
4. **Evidence-Based Policy Development**:
   - Policymakers should rely on robust clinical and economic evidence to guide decisions about wearable technology regulation and reimbursement.

1.12 Conclusion

Wearable technologies represent a paradigm shift in healthcare delivery, offering unprecedented opportunities for personalized care, preventive health, and cost savings. However, their success depends on the development of comprehensive health policies and reimbursement frameworks that address regulatory challenges, incentivize adoption, and ensure equitable access. By prioritizing innovation, interoperability, and patient-centered care, policymakers and stakeholders can harness the full potential of wearable technologies to transform healthcare systems and improve outcomes for patients worldwide.

References

1. Jafleh, E. A., Alnaqbi, F. A., Almaeeni, H. A., Faqeeh, S., Alzaabi, M. A., & Al Zaman, K.: The role of wearable devices in chronic disease monitoring and patient care: A comprehensive review. Cureus, 16(9), e68921 (2024).
2. Rutherford, J.J.: Wearable technology. IEEE Eng. Med. Biol. Mag. **29**(3), 19–24 (2010)
3. Nkurunziza, J.M.V., Udahemuka, J.C., Dusenge, J.B., Umutesi, F.: Overview of trending medical technologies. Glob. Clin. Eng. J. **4**(3), 16–46 (2022)
4. Velasquez, G.D.D.S., Borna, S., Maniaci, M.J., Coffey, J.D., Haider, C.R., Demaerschalk, B.M., Forte, A.J.: Economic perspective of the use of wearables in health care: a systematic review. Mayo Clinic Proc. Digit. Health **2**(3), 299–317 (2024)
5. Lu, L., Zhang, J., Xie, Y., et al.: Wearable health devices in health care: narrative systematic review. JMIR Mhealth Uhealth **8**(11), e18907 (2020). https://doi.org/10.2196/18907
6. Fuller, D., Colwell, E., Low, J., Orychock, K., Tobin, M.A., Simango, B., Taylor, N.G.: Reliability and validity of commercially available wearable devices for measuring steps, energy expenditure, and heart rate: systematic review. JMIR mHealth and uHealth **8**(9), e1869 (2020)
7. Soliño-Fernandez, D., Ding, A., Bayro-Kaiser, E., Ding, E.L.: Willingness to adopt wearable devices with behavioral and economic incentives by health insurance wellness programs: results

of a US cross-sectional survey with multiple consumer health vignettes. BMC Public Health **19**(1), 1649 (2019)

8. Eisner, L.: Wearables for medical and wellness uses: standards and FDA guidance documents. In Compliance. https://incompliancemag.com/wearables-for-medical-and-wellness-uses-standards-and-fda-guidance-documents/ (2017, May 17). Last accessed 2024/12/14
9. Mattison, G., Canfell, O., Forrester, D., Dobbins, C., Smith, D., Töyräs, J., Sullivan, C.: The influence of wearables on health care outcomes in chronic disease: systematic review. J. Med. Internet Res. **24**(7), e36690 (2022)
10. Kim, H.J., Lee, K.H., Lee, J.H., Youk, H., Lee, H.Y.: The effect of a mobile and wearable device intervention on increased physical activity to prevent metabolic syndrome: observational study. JMIR Mhealth Uhealth **10**(2), e34059 (2022)
11. Wall, C., Hetherington, V., Godfrey, A.: Beyond the clinic: the rise of wearables and smartphones in decentralising healthcare. NPJ Digit. Med. **6**, 219 (2023)
12. Kooiman, T.J., Dontje, M.L., Sprenger, S.R., Krijnen, W.P., Van der Schans, C.P., de Groot, M.: Reliability and validity of ten consumer activity trackers. BMC Sports Sci. Med. Rehabil. **7**, 1–11 (2015)
13. Strain, T., Wijndaele, K., Dempsey, P.C., Sharp, S.J., Pearce, M., Jeon, J., Brage, S.: Wearable-device-measured physical activity and future health risk. Nat. Med. **26**(9), 1385–1391 (2020)
14. Davidson, K.P.: Breathe through your nose, don't pay through it: The impact the healthcare industry has on nasal function and how we breathe. PeeDee Publishing (2022)
15. Weng, J.K., Virk, R., Kaiser, K., Hoffman, K.E., Goodman, C.R., Mitchell, M., Smith, B.D.: Automated, real-time integration of biometric data from wearable devices with electronic medical records: a feasibility study. JCO Clin. Cancer Inf. **8**, e2400040 (2024)
16. van Westrhenen, A., Lazeron, R.H., van Dijk, J.P., Leijten, F.S., Thijs, R.D., Dutch TeleEpilepsy Consortium: Multimodal nocturnal seizure detection in children with epilepsy: a prospective, multicenter, long-term, in-home trial. Epilepsia **64**(8), 2137–2152 (2023)
17. Nherera, L., Larson, B., Cooley, A., Reinhard, P.: An economic analysis of a wearable patient sensor for preventing hospital-acquired pressure injuries among the acutely ill patients. Int. J. Health Econ. Manage. **21**(4), 457–471 (2021)
18. Diemberger, I., Gardini, B., Martignani, C., Ziacchi, M., Corzani, A., Biffi, M., Boriani, G.: Holter ECG for pacemaker/defibrillator carriers: what is its role in the era of remote monitoring? Heart **101**(16), 1272–1278 (2015)
19. Jain, P.K., Tiwari, A.K.: Heart monitoring systems—a review. Comput. Biol. Med. **54**, 1–1 (2014)
20. Oseran, A.S., Ati, S., Feldman, W.B., Gondi, S., Yeh, R.W., Wadhera, R.K.: Assessment of prices for cardiovascular tests and procedures at top-ranked US hospitals. JAMA Intern. Med. **182**(9), 996–999 (2022)
21. Kinlay, S., Leitch, J.W., Neil, A., Chapman, B.L., Hardy, D.B., Fletcher, P.J.: Cardiac event recorders yield more diagnoses and are more cost-effective than 48-hour Holter monitoring in patients with palpitations: a controlled clinical trial. Ann. Internal Med. **124**(1_Part_1), 16–20 (1996)
22. Orlov, L.M.: The future of wearables and older adults 2021. aging and health technology watch (June 2021). https://www.ageinplacetech.com/page/future-wearables-and-olderadults-2021
23. Shamim-Uzzaman, Q.A., Bae, C.J., Ehsan, Z., Setty, A.R., Devine, M., Dhankikar, S., Epstein, L.J.: The use of telemedicine for the diagnosis and treatment of sleep disorders: an American Academy of Sleep Medicine update. J. Clin. Sleep Med. **17**(5), 1103–1107 (2021)
24. Mulgund, P., Sharman, R., Rifkin, D., Marrazzo, S.: Design, development, and evaluation of a telemedicine platform for patients with sleep apnea (Ognomy): design science research approach. JMIR Formative Res. **5**(7), e26059 (2021)
25. Braun, M., Stockhoff, M., Tijssen, M., Dietz-Terjung, S., Coughlin, S., Schöbel, C.: A systematic review on the technical feasibility of home-polysomnography for diagnosis of sleep disorders in adults. Curr. Sleep Med. Rep., 1–13 (2024)
26. Deutsch, P.A., Simmons, M.S., Wallace, J.M.: Cost-effectiveness of split-night polysomnography and home studies in the evaluation of obstructive sleep apnea syndrome. J. Clin. Sleep Med. **2**(02), 145–153 (2006)

27. Quaresima, V., Ferrari, M., Scholkmann, F.: Ninety years of pulse oximetry: history, current status, and outlook. J. Biomed. Opt. **29**(S3), S33307–S33307 (2024)
28. Wickwire, E.M., Verma, T.: Value and payment in sleep medicine. J. Clin. Sleep Med.Clin. Sleep Med. **14**(5), 881–884 (2018)
29. Taylor, M.L., Thomas, E.E., Snoswell, C.L., Smith, A.C., Caffery, L.J.: Does remote patient monitoring reduce acute care use? A systematic review. BMJ Open **11**(3), e040232 (2021)
30. Dickinson, M.G., Allen, L.A., Albert, N.A., DiSalvo, T., Ewald, G.A., Vest, A.R., Givertz, M.M.: Remote monitoring of patients with heart failure: a white paper from the Heart Failure Society of America Scientific Statements Committee. J. Cardiac Failure **24**(10), 682–694 (2018)
31. Kouaho, W.J., Epstein, D.A.: Investigating Perspectives of and Experiences with Low Cost Commercial Fitness Wearables. Proceedings of the ACM on Interactive, Mobile, Wearable and Ubiquitous Technologies **8**(4), 1–22 (2024)
32. Nunn, R., Parsons, J., Shambaugh, J.: A dozen facts about the economics of the US health-care system. The Hamilton Project, Economic Facts (2020)
33. De Guzman, K.R., Snoswell, C.L., Taylor, M.L., Gray, L.C., Caffery, L.J.: Economic evaluations of remote patient monitoring for chronic disease: a systematic review. Value in Health **25**(6), 897–913 (2022)

Dr. Karen Parker Davidson, RN is an international expert in airway health and clinical nurse specialist with over 30 years of experience. She pioneered nasal airflow diagnostics and introduced several tools, including the Davidson Maneuver and the MAVERICK Nasal Exam™. Dr. Davidson is the founder of Airway AI and serves as VP at GM Instruments. She has extensive experience in critical care, oncology, and flight nursing. Holding a DHA and three master's degrees, she is a faculty member at Liberty University and Central Michigan University, and recipient of multiple awards including the 2024 NSF Award.

Connected Health: Transforming Remote Patient Care with Wearable Device Applications

Wasswa Shafik

Abstract Connected health is revolutionizing remote patient care through the integration of wearable device applications, enabling continuous monitoring, real-time data collection, and enhanced patient engagement. This chapter explores the transformative role of wearable technologies in bridging gaps between traditional healthcare models and patient-centered digital solutions. By leveraging biosensors, artificial intelligence (AI), and cloud-based analytics, wearable devices provide clinicians with actionable insights, allowing for early disease detection, personalized treatment adjustments, and improved chronic disease management. The chapter critically examines the technological infrastructure of connected health, including data interoperability, cybersecurity concerns, and regulatory frameworks that govern wearable medical devices. Additionally, it assesses the impact of these innovations on health outcomes, cost efficiency, and healthcare accessibility, particularly in underserved populations. The discussion extends to the role of machine learning algorithms in predictive analytics, facilitating proactive interventions in conditions such as cardiovascular disease, diabetes, and respiratory disorders. As healthcare systems shift toward value-based care, wearable devices emerge as a pivotal tool in reducing hospitalizations, optimizing resource allocation, and fostering patient autonomy. This chapter concludes by addressing current challenges, ethical considerations, and future directions for wearable technologies in remote patient care, emphasizing their potential to reshape healthcare delivery in an increasingly digital world.

Keywords Wearable devices · Remote patient care · Disease management · Telehealth · Monitoring · Innovation

W. Shafik (✉)
Dig Connectivity Research Laboratory (DCRLab), Kampala, Uganda
e-mail: wasswashafik@ieee.org

School of Digital Science, Universiti Brunei Darussalam, Gadong, Brunei

P. Eappen et al. (eds.), *Remote Monitoring and Wearable Devices in Healthcare*, Information Systems Engineering and Management 63,
https://doi.org/10.1007/978-3-031-98897-4_2

1 Introduction

Connected health, a model of healthcare delivery supported by technology and data sharing, is reshaping how patients interact with providers. Wearable devices are at the forefront of this transformation, enabling continuous health monitoring and empowering individuals to take an active role in their well-being. From fitness trackers to medical-grade devices, wearables have evolved into powerful tools for managing chronic conditions, predicting health risks, and facilitating personalized medicine. This chapter delves into the critical role of wearable devices in connected health, highlighting their impact on patient care and the healthcare system at large.

Connected health is revolutionizing remote patient care by integrating wearable device applications, enabling continuous monitoring and real-time data collection [1]. These advancements enhance patient engagement and provide clinicians with actionable insights, facilitating early disease detection and personalized treatment adjustments and monitoring [2, pp. 102–106]. Wearable technologies, such as biosensors and digital biomarkers, play a pivotal role in bridging gaps between traditional healthcare models and patient-centered digital solutions [3].

Furthermore, clinical research is being transformed through the use of telemedicine, mobile healthcare providers, and remove data collections, which was only accelerated by the COVID-19 pandemic. Connected health allows remote data from clinical trial participants to be electronically transmitted [4]. Data such as laboratory results, safety information, and outcome measures reported by patients, clinicians, or observers, which marks a significant advancement in obtaining this level of data [3].

One survey found that 67% of respondents had less than 25% of their clinical trials (CTs) decentralized, and none had fully decentralized trials. Only 50% had prior experience with remote data capture (RDC), wearables, or biomarkers, mainly in CTs, while 40% had implemented RDC in observational studies. Wearables were the most commonly used RDC method (59%), with wrist-worn devices being the most frequent [3].

Wearable health monitoring devices and digital biomarkers play a crucial role in this shift, enabling continuous measurement of physiological parameters outside traditional clinical settings [5]. These technologies enhance patient care and biomedical research by facilitating remote monitoring and improving DCT efficiency [3].

While remote data capture, wearables, and digital biomarkers offer many benefits, challenges remain. To assess the adoption and obstacles of these technologies in DCTs, the authors conducted an online survey of clinical research stakeholders in India, analyzing 80 responses. This study highlights the growing role of digital tools in clinical research while acknowledging the need to address existing challenges for broader implementation [3].

The technological infrastructure of connected health encompasses data interoperability, cybersecurity, and regulatory frameworks governing wearable medical devices. Addressing these aspects ensures the seamless integration of wearables into healthcare systems, promoting improved health outcomes, cost efficiency, and accessibility, particularly in underserved populations [6].

As healthcare systems shift toward value-based care, wearable devices emerge as pivotal tools in reducing hospitalizations, optimizing resource allocation, and fostering patient autonomy. However, challenges such as data privacy, user adherence, and the need for standardized protocols persist. Ongoing research and development aim to overcome these hurdles, emphasizing the potential of wearable technologies to reshape healthcare delivery in an increasingly digital world [7].

1.1 Development and Evolution of Connected Health and Wearable Devices

The journey of connected health and remote patient care reflects a progression from early telecommunication methods to sophisticated, technology-driven solutions. The integration of wearable devices has been pivotal in this evolution, offering new avenues for patient engagement and proactive healthcare management.

Early Developments in Remote Patient Care

The concept of remote patient care has evolved significantly over the past century, but its origins can be traced back to the late 19th and early twentieth centuries, when advancements in communication technology first enabled healthcare to extend beyond traditional clinical settings. While modern telemedicine relies on digital platforms, cloud computing, and artificial intelligence, the foundation of remote patient care was laid through early innovations aimed at bridging the gap between patients and medical professionals over long distances [8].

One of the earliest instances of remote healthcare can be seen in the late 1800s, when physicians began using telegraphy to communicate medical information. The ability to transmit messages over long distances allowed doctors to consult with colleagues, relay diagnoses, and even guide treatments for patients in remote areas [9]. This marked the first step toward leveraging technology to overcome geographical barriers in healthcare delivery.

By the early twentieth century, the telephone revolutionized remote healthcare by enabling direct communication between doctors and patients. Physicians could provide medical advice, assess symptoms, and even prescribe treatments without requiring patients to visit a hospital or clinic. This was particularly valuable in rural communities where access to healthcare facilities was limited. Radio technology further expanded remote patient care, especially in the 1920s and 1930s, when medical professionals began using radio transmissions to provide healthcare consultations to patients in isolated locations [10]. One notable example is the Royal Flying Doctor Service in Australia, established in 1928, which used two-way radios to offer medical guidance to individuals living in remote regions.

The mid-twentieth century saw the introduction of early telemedicine initiatives, driven by advancements in television and video transmission. In the 1950s and 1960s, experiments in remote healthcare began integrating real-time audio-visual communication. One of the pioneering projects was a collaboration between the Nebraska Psychiatric Institute and Norfolk State Hospital in the 1950s, where closed-circuit television was used for psychiatric consultations between physicians and patients in separate locations. Around the same time, National Aeronautics and Space Administration (NASA) played a crucial role in advancing remote patient monitoring technologies. As space exploration progressed, the need to monitor astronauts' physiological parameters in real time led to the development of early biosensors, which could track vital signs such as heart rate, respiration, and body temperature from a distance. These innovations laid the groundwork for modern wearable health technologies.

By the 1970s and 1980s, telemedicine expanded further with the emergence of satellite-based communication systems. These advancements allowed for more sophisticated remote healthcare services, including teleradiology, where medical images such as X-rays and CT scans could be transmitted across vast distances for expert review. Early applications of electronic health records (EHRs) and digital data sharing also began to take shape, setting the stage for today's connected health infrastructure.

The foundations established by these early developments in remote patient care have paved the way for modern telemedicine, wearable health technologies, and AI-driven diagnostics. The seamless data exchange seen today between patients and healthcare providers—enabled by cloud computing, mobile applications, and real-time remote monitoring—has its roots in these historical innovations [11]. As technology continues to advance, the principles of early remote patient care remain central to the ongoing transformation of global healthcare, making high-quality medical services more accessible, efficient, and personalized than ever before.

The foundations of remote patient care date back to the late 19th and early twentieth centuries, when communication technologies began to play a role in medicine. While modern telemedicine relies on digital connectivity, its roots lie in early innovations that sought to extend healthcare beyond traditional clinical settings.

Telemedicine in the 19th and Early 20th Centuries

The 19th-century introduction of the telegraph facilitated the transmission of medical consultations over long distances, particularly in military and maritime settings. By the early twentieth century, radio communication enabled ship-to-shore medical advice, significantly aiding seafarers without onboard physicians. The advent of the telephone further refined telemedicine, allowing real-time physician–patient interactions, albeit limited by infrastructure constraints. Early applications were largely consultative, focused on relaying symptoms and receiving diagnostic guidance rather than direct intervention [12]. These developments laid the foundation for modern telemedicine by demonstrating the feasibility of remote medical assistance, despite technological limitations that restricted its scope to primarily advisory roles. By 1906, electrocardiograms (ECGs) were transmitted over telephone lines for the first time by Dutch physiologist Willem Einthoven [13]. This marked a significant breakthrough, as it demonstrated the feasibility of remotely monitoring a patient's heart activity.

The Role of Radio and Military Applications

The invention and widespread adoption of radio communication in the early twentieth century further advanced remote healthcare. In 1925, *Science and Invention* magazine published an article envisioning a device called the "teledactyl," which would allow doctors to see and examine patients remotely via radio and mechanical arms [14]. While purely conceptual at the time, this idea foreshadowed modern telehealth solutions. During World War II, remote medical consultations became a necessity for soldiers and medical personnel stationed in distant locations [15]. Military telemedicine initiatives focused on using radio and early computing technologies to assist with diagnosing and treating wounded soldiers in battlefield settings.

NASA and the Expansion of Remote Monitoring

During the 1960s, the U.S. space program played a transformative role in the development of remote patient monitoring (RPM) technologies, as NASA sought innovative solutions to track astronauts' health in real time during space missions. The extreme and isolated conditions of space travel made it imperative to develop a system that could continuously monitor physiological parameters such as heart rate, respiration, body temperature, and blood pressure without requiring direct medical intervention. This necessity led to groundbreaking advancements in biomedical telemetry, setting the stage for the future of telemedicine and remote patient care.

In the 1960s, NASA pioneered remote patient care as part of its efforts to monitor astronauts' vital signs during space missions [16]. A major milestone in this effort was the development of the Integrated Medical and Behavioral Laboratory Measurement System (IMBLMS) by NASA in collaboration with researchers from the aerospace and medical fields. The Integrated Medical and Behavioral Laboratory Measurement System (IMBLMS) was developed to track astronauts' health in real time while they were in space [17]. This research contributed significantly to the advancement of remote patient monitoring (RPM) technologies that would later be adopted in hospitals and home healthcare.

In more detail, the IMBLMS was designed to collect, analyze, and transmit real-time health data from astronauts to mission control on Earth, ensuring that medical teams could detect early signs of physiological stress, dehydration, cardiovascular irregularities, or other health concerns. This system relied on non-invasive biosensors placed on astronauts' bodies, which continuously transmitted health metrics via radio telemetry. The ability to remotely monitor astronauts' health in microgravity environments without direct physician access was a revolutionary achievement that laid the groundwork for future RPM applications in civilian healthcare.

The innovations emerging from NASA's space medicine research had far-reaching implications beyond spaceflight. As the technology evolved, hospitals and healthcare institutions began exploring how these advanced telemetry systems could be adapted for patient monitoring on Earth. The principles behind IMBLMS were later applied to intensive care units (ICUs), where continuous monitoring of critically ill patients became standard practice. This technology also influenced the development of early portable electrocardiogram (ECG) monitors, which allowed for remote heart monitoring in patients with cardiovascular diseases. The ability to collect real-time physiological data from patients outside of clinical settings, a necessity for astronauts in space, became a cornerstone of modern home healthcare and chronic disease management.

NASA's pioneering research in biomedical telemetry also contributed to the advancement of wearable health technology. The miniaturization of sensors and the ability to transmit health data wirelessly led to the development of early Holter monitors, which became widely used for ambulatory ECG monitoring in cardiac patients. This technology eventually evolved into modern wearable health devices, such as smartwatches and fitness trackers, which now integrate remote monitoring capabilities for conditions like arrhythmias, diabetes, and sleep disorders.

Furthermore, NASA's advancements in remote healthcare solutions influenced the early development of telemedicine programs, particularly for rural and underserved populations. The ability to monitor astronauts remotely demonstrated that high-quality medical care could be delivered without requiring a patient's physical presence in a hospital. This concept was later expanded into telehealth initiatives, where satellite-based communication and remote diagnostics were used to provide healthcare access to remote locations, military bases, and even disaster-stricken areas.

The legacy of NASA's 1960s research in remote patient monitoring is evident in today's healthcare landscape. From hospital-based telemetry systems to AI-driven home health monitoring, the foundational work done to ensure astronaut health in space has played a critical role in shaping the evolution of modern connected health technologies. As wearable devices and artificial intelligence continue to advance, the principles established by NASA's early remote monitoring systems remain fundamental to the future of personalized, data-driven, and accessible healthcare worldwide.

Emergence of Hospital-Based Telemedicine

While NASA was pioneering remote patient monitoring in space, hospitals and research institutions were simultaneously exploring ways to use emerging communication technologies to expand healthcare access on Earth. One of the earliest and most notable telemedicine initiatives was launched in 1967 by the University of Nebraska Medical Center (UNMC) [18]. The initiative demonstrated the practical use of telehealth for improving access to specialized care, especially in rural areas.

This groundbreaking program used two-way video communication to facilitate remote psychiatric consultations between doctors at UNMC and patients located at Norfolk State Hospital, a psychiatric institute situated 112 miles away. This innovative approach demonstrated the practical application of telehealth in connecting patients with specialists, particularly in underserved and rural communities where access to specialized healthcare was limited.

At the time, psychiatric care in remote areas faced significant challenges. Many patients in rural regions lacked access to mental health professionals, and those requiring specialized psychiatric evaluations or therapy often had to travel long distances to urban medical centers. The telemedicine system developed by UNMC addressed this issue by allowing psychiatrists to conduct real-time evaluations and consultations via a closed-circuit television system. This program enabled doctors to diagnose mental health conditions, offer treatment recommendations, and provide therapy sessions without requiring patients to leave their local facilities.

The success of this early telemedicine experiment provided compelling evidence that remote healthcare was not only feasible but also highly effective in reducing barriers to specialized care. It demonstrated that real-time video communication could facilitate clinical decision-making, enhance doctor-patient interactions, and expand access to healthcare services beyond traditional hospital settings. The initiative also proved valuable for reducing patient transfers, as many individuals could receive necessary psychiatric care remotely rather than being transported to distant medical centers, which reduced costs and improved continuity of care.

Beyond psychiatric care, the UNMC telemedicine project set the stage for the broader adoption of telehealth technologies across different medical fields. As video communication technology improved, telemedicine expanded to other specialties, including radiology, cardiology, dermatology, and primary care. For example, radiologists began using early forms of teleradiology, where X-rays and CT scans could be electronically transmitted for remote interpretation, eliminating the need for physical film transport and accelerating diagnostic processes.

The University of Nebraska's pioneering work also influenced future developments in rural healthcare programs, veterans' healthcare, and disaster response telemedicine. By demonstrating the effectiveness of remote consultations, this early initiative laid the groundwork for subsequent government and healthcare system investments in telehealth infrastructure, particularly for communities that faced geographical barriers to medical services. Over the following decades, healthcare organizations and policymakers built upon this model, leading to the expansion of Medicare reimbursements for telehealth, federal telemedicine grants, and large-scale implementation of remote healthcare networks.

Today, the principles established by the 1967 University of Nebraska telemedicine program continue to drive the evolution of digital health. The fundamental idea of delivering care remotely through live video consultations has evolved into modern telehealth services powered by AI-driven diagnostics, mobile health apps, and wearable health technologies. The early adoption of telemedicine in psychiatric care also foreshadowed today's widespread use of virtual mental health platforms, cognitive behavioral therapy apps, and remote psychiatric interventions, which have become critical tools in managing mental health crises and expanding access to care.

Ultimately, the 1967 UNMC telemedicine initiative was a defining moment in the history of connected health, proving that technology could break down barriers to medical care, improve patient outcomes, and lay the foundation for the telemedicine revolution that continues to shape modern healthcare delivery.

The Digital Revolution and Connected Health Beginnings

By the 1980s and 1990s, advancements in computer networks, satellite communications, and digital imaging facilitated the growth of telemedicine and remote patient care [19]. The rise of electronic health records (EHRs) and the increasing availability of home-based monitoring devices allowed healthcare providers to track patients' health outside of clinical environments.

These early developments laid the groundwork for today's connected health ecosystem, where wearable devices, mobile health apps, and AI-powered analytics are transforming patient care. The increasing reliance on real-time data, remote diagnostics, and continuous monitoring has made remote patient care more accessible and effective than ever before.

1.2 Technological Advancements in Wearable Medical Devices for Connected Health

The rapid evolution of sensors, miniaturization, and wireless connectivity has revolutionized the capabilities of wearable devices in healthcare. Once primarily associated with fitness tracking, wearables have now expanded into the medical domain, offering real-time health monitoring, remote diagnostics, and early disease detection. These advancements have significantly improved patient outcomes by allowing continuous data collection, reducing hospital visits, and enabling proactive medical interventions.

Miniaturization and Sensor Technology

One of the most significant breakthroughs in wearable healthcare technology is sensor miniaturization [20]. Modern sensors are now smaller, more energy-efficient, and highly sensitive, allowing them to be embedded in compact wearable devices such as smartwatches, patches, and even smart textiles. Advanced biosensors can measure a variety of physiological parameters, including heart rate, respiratory rate, oxygen saturation (SpO_2), blood pressure, body temperature, and glucose levels. Some cutting-edge wearables even integrate electrocardiography (ECG) and photoplethysmography (PPG) sensors to provide deeper cardiovascular insights.

Glucose monitoring is a prime example of how sensor technology has transformed chronic disease management. Traditional blood glucose monitoring required finger-prick tests, which could be painful and inconvenient. However, continuous glucose monitoring (CGM) devices, equipped with minimally invasive sensors, now provide real-time glucose readings, allowing diabetic patients to better manage their condition with fewer disruptions to their daily lives.

Wireless Connectivity and Real-Time Data Transmission

The integration of wireless connectivity technologies such as Bluetooth, Wi-Fi, and 5G has further enhanced the functionality of wearable medical devices [21]. These technologies enable seamless communication between wearables and external systems, allowing healthcare providers to remotely monitor patients in real-time. With cloud-based platforms and AI-driven analytics, medical professionals can access continuous streams of patient data, detect anomalies, and intervene before conditions worsen.

For instance, wearable ECG monitors can detect arrhythmias or atrial fibrillation and send alerts to a physician instantly, reducing the risk of stroke or cardiac arrest [22]. Similarly, pulse oximeters embedded in smartwatches can track oxygen saturation levels, which became particularly crucial during the COVID-19 pandemic for monitoring respiratory health.

AI Integration and Predictive Healthcare

Beyond simple monitoring, modern wearables leverage artificial intelligence (AI) and machine learning algorithms to analyze data and provide predictive insights. AI-powered wearables can detect subtle changes in physiological patterns, predicting potential health issues before symptoms appear [23]. For example, some smartwatches are now capable of identifying early signs of atrial fibrillation, prompting users to seek medical attention before a serious event occurs.

Moreover, AI-driven personalized health recommendations help users adopt healthier behaviors. As mentioned throughout the textbook in associated chapters, wearables can track sleep patterns, stress levels, and activity trends, offering tailored advice on exercise, nutrition, and recovery. These proactive interventions contribute to long-term health improvement and disease prevention.

@@**The Future of Wearable Medical Technology.**

As technology continues to evolve, the future of wearable medical devices looks promising. Advancements in nanotechnology, biocompatible materials, and flexible electronics will lead to even more sophisticated wearables, capable of non-invasive blood pressure monitoring, hydration assessment, and early cancer detection. Additionally, implantable and bio-integrated sensors may further bridge the gap between wearable technology and traditional medicine, enabling even more precise, continuous, and unobtrusive health monitoring.

1.3 Integration with Connected Health: Transforming Remote Patient Care with Wearable Devices

Currently, 76% of hospitals in the U.S. utilize telehealth services to expand patient care. Furthermore, the integration of wearable devices with connected health platforms has revolutionized healthcare delivery, shifting from traditional, episodic care models to a continuous, data-driven, and patient-centric approach.

Wearables play a critical role in connected health by facilitating seamless data transmission between patients, healthcare providers, and digital health ecosystems. These devices continuously collect biometric data, including heart rate, blood pressure, oxygen saturation, and activity levels, transmitting this information to cloud-based platforms. AI-driven analytics then process this data in real time, alerting both patients and clinicians to any concerning trends or anomalies. This proactive approach allows for timely medical intervention, reducing hospital admissions and improving overall patient outcomes.

Telemedicine has further benefited from wearable integration by enabling virtual consultations based on real-time health metrics. Physicians can review continuously updated patient data, offering personalized recommendations without requiring in-person visits. This has proven particularly beneficial for managing chronic diseases, elder care, and rural healthcare access, where distance and mobility limitations often pose challenges. Additionally, wearables equipped with geolocation tracking and fall detection capabilities support aging populations, ensuring rapid emergency response when needed.

The ability of wearables to synchronize with electronic health records (EHRs) has also streamlined healthcare workflows, reducing administrative burdens and enhancing patient monitoring efficiency. As these technologies continue to advance, they will play an even greater role in predictive analytics, enabling early diagnosis and individualized treatment plans tailored to a patient's unique physiological patterns.

Despite these advantages, challenges remain in integrating wearables with connected health platforms. Data security, interoperability between different healthcare systems, and patient privacy concerns must be addressed to maximize the benefits of these technologies. Regulatory frameworks and standardization efforts are crucial in ensuring that wearable-derived data is both reliable and actionable within clinical settings.

By seamlessly exchanging real-time health data, these platforms enable remote monitoring, telemedicine consultations, and evidence-based clinical decision-making. This fusion of wearable devices and connected health solutions is, making wearables an essential component of modern healthcare ecosystems. As technology continues to evolve, the adoption of wearables in healthcare ecosystems will further enhance patient engagement, improve health outcomes, and redefine the future of medical care delivery.

Seamless Data Exchange and Interoperability

Seamless data exchange and interoperability represent foundational pillars in the evolution of connected health, offering a paradigm shift in the way healthcare data is captured, shared, and analyzed. In an interconnected healthcare ecosystem, the capacity for wearable devices to effortlessly communicate with electronic health records (EHRs), cloud-based analytics, and artificial intelligence (AI)-driven diagnostic systems is a key advancement. Interoperability ensures that disparate healthcare systems and devices can work together, facilitating a holistic view of a patient's health.

Wearable Devices and Real-Time Data Collection

Wearable devices are increasingly integral in capturing a broad array of physiological data. These devices, which include smartwatches, fitness trackers, and medical-grade wearables, collect continuous measurements such as heart rate, oxygen saturation levels, glucose concentrations, respiratory rate, and sleep patterns. The data is transmitted wirelessly via technologies such as Bluetooth, Wi-Fi, and 5G networks to secure, cloud-based digital health platforms. This seamless data transfer allows for real-time monitoring of patient health, a marked departure from the traditional reliance on periodic in-person visits.

For instance, a wearable ECG-enabled smartwatch might monitor a patient's heart rhythm continuously. As the data is gathered, it is automatically uploaded to the patient's healthcare provider's dashboard, giving the clinician an up-to-the-minute view of the patient's cardiovascular status. This ongoing flow of information ensures that any deviations from the patient's baseline health metrics are immediately detected.

Enhanced Decision-Making Through Integrated Data

With the advent of interoperability, data from multiple sources can be synthesized to provide a comprehensive picture of the patient's health. Cloud-based platforms serve as centralized hubs where wearable device data is integrated with historical medical records, lab results, imaging studies, and more. This enables healthcare providers to make informed, evidence-based decisions without relying on incomplete or outdated information.

Moreover, AI-driven tools can analyze this integrated data in real time, offering predictive insights, flagging potential risks, and suggesting personalized interventions. For example, an AI system might recognize an emerging abnormality in heart rate variability, prompting an immediate recommendation for additional diagnostic testing or an adjustment to the patient's treatment regimen.

1.4 Applications of Wearable Devices in Remote Patient Care

Chronic Disease Management

Chronic disease management represents one of the most promising areas where connected health technologies, particularly wearable devices, can have a profound impact. Chronic diseases—conditions such as cardiovascular disease, diabetes, hypertension, respiratory disorders, and other long-term health issues—often require continuous or frequent monitoring to prevent complications and optimize treatment. Traditionally, this has involved periodic in-person visits to healthcare providers for routine checks, which can be cumbersome for patients and inefficient for the healthcare system. The integration of wearable devices into this process, however, is shifting the paradigm from reactive to proactive care, facilitating a more continuous and real-time approach to managing chronic conditions.

Remote monitoring by entering or capturing data has been popular in the home-health specialty. These platforms have developed into various technologies as the predecessor to wearable technologies. Wearable devices have become essential tools in managing chronic diseases such as diabetes, hypertension, and respiratory conditions [24]. Continuous glucose monitors (CGMs) provide real-time blood sugar readings, enabling diabetic patients to make informed decisions about insulin use, diet, and activity levels [25]. These devices also feature alerts that warn users of dangerously high or low glucose levels, preventing severe complications like diabetic ketoacidosis or hypoglycemic episodes.

Smart blood pressure monitors allow hypertensive patients to track fluctuations in blood pressure throughout the day, providing critical data that can help healthcare providers adjust medications and lifestyle recommendations. Similarly, wearable spirometers assist patients with respiratory diseases such as chronic obstructive pulmonary disease (COPD) and asthma by monitoring lung function and detecting early signs of exacerbation. Some advanced models even integrate with AI-powered analytics to provide predictive insights on respiratory health.

By integrating with mobile applications and cloud-based health platforms, these wearables support personalized disease management, reduce emergency hospital visits, and enhance patient engagement in their own healthcare. The ability to generate long-term data trends also allows physicians to assess the effectiveness of treatment plans and make data-driven adjustments.

The Power of Continuous Monitoring

One of the key advantages of wearable devices is their ability to provide continuous monitoring of vital health metrics, such as heart rate, blood pressure, glucose levels, oxygen saturation, and respiratory rate. Unlike traditional healthcare monitoring methods, which are often limited to discrete, episodic check-ups or tests, wearables collect data in real time and allow for ongoing surveillance of a patient's health status. This constant data stream is transmitted securely to cloud-based health platforms or directly to healthcare providers, where it can be analyzed and used to inform clinical decisions. For patients with chronic conditions, this means that their health can be monitored 24/7 without the need to schedule and attend frequent in-person appointments. Conditions particularly suited to continuous monitoring are diabetes, hypertension, and respiratory disorders to name a few.

Diabetes Management through Wearable Technology

In diabetes management, wearable devices such as continuous glucose monitors (CGMs) are transforming care by allowing for constant surveillance of glucose levels. The data from CGMs is typically transmitted via Bluetooth to a patient's smartphone or directly to a cloud-based platform accessible by their healthcare provider. This real-time access to glucose data means that healthcare providers can make adjustments to treatment plans based on immediate feedback. For example, if a patient's glucose levels are trending too high, the healthcare team can modify their insulin dosage or recommend lifestyle adjustments such as changes to diet or exercise.

For example, a patient managing diabetes can wear a continuous glucose monitor (CGM) that tracks blood sugar levels throughout the day. This real-time data, which is automatically uploaded to the patient's healthcare team, allows clinicians to identify fluctuations or trends that might otherwise go unnoticed. If a patient's glucose levels rise unexpectedly, a clinician can intervene before the patient experiences significant complications, such as diabetic ketoacidosis or hypoglycemic episodes.

This real-time data allows clinicians to intervene earlier, preventing complications before they arise. In addition to clinical interventions, continuous monitoring can empower patients to take a more active role in their care. By receiving timely feedback on how their lifestyle choices affect their glucose levels, patients can make more informed decisions about their daily habits and work with their care team to optimize their health outcomes. The data collected from CGMs also helps to identify patterns that may not be apparent during infrequent clinic visits, providing a more nuanced view of the patient's disease progression and response to treatment.

Cardiovascular Disorders and Continuous Monitoring

Another example of how wearable devices can enhance chronic disease management is in the monitoring of hypertension. Hypertension, or high blood pressure, is a common and often asymptomatic condition that can lead to serious complications such as stroke, heart disease, and kidney damage if left untreated. Traditionally, blood pressure is measured during a clinical visit, and a patient's treatment plan is adjusted based on those isolated readings. However, because blood pressure can fluctuate throughout the day, relying on a single reading during a clinic visit may not provide an accurate representation of a patient's true blood pressure status.

Wearable blood pressure monitors, which continuously track and transmit blood pressure data, offer a more comprehensive view of a patient's cardiovascular health. These devices, which can be worn on the wrist or upper arm, provide data that is uploaded to a healthcare provider's system for ongoing analysis. This allows clinicians to monitor trends in a patient's blood pressure over time and make timely adjustments to medication or lifestyle recommendations. Suppose a patient's blood pressure readings indicate that their treatment is no longer effective. In that case, healthcare providers can respond immediately, ensuring that the patient's condition is managed before any serious complications occur.

Moreover, wearable blood pressure monitors can help address the phenomenon of "white coat syndrome," where a patient's blood pressure spikes due to the stress of a clinic visit. By capturing data in a variety of real-life settings, wearables provide a more accurate picture of a patient's blood pressure under normal conditions, leading to more precise treatment plans. Cardiovascular disease is another area where connected health technologies are making significant strides.

Conditions such as atrial fibrillation (AFib), arrhythmias, and heart failure require continuous monitoring to detect early signs of deterioration and prevent potentially life-threatening events. Wearable ECG (electrocardiogram) devices, such as those integrated into smartwatches, allow patients to track their heart rhythms continuously. These devices can automatically upload data to a healthcare provider's dashboard, enabling real-time analysis by physicians. Many of these technologies are discussed throughout the textbook, but here we examine the connected to health and health systems.

For instance, if a patient with known atrial fibrillation experiences irregular heartbeats, their wearable ECG monitor will capture and transmit this data to their healthcare team, triggering an alert. The clinician can then review the data, confirm the diagnosis, and recommend an intervention, such as adjusting medication or scheduling a follow-up appointment. This early detection capability significantly reduces the risk of serious complications, such as stroke or heart failure exacerbations, by ensuring that patients receive timely and appropriate care.

Respiratory Disorders and Remote Monitoring of Oxygen Saturation

Wearable devices also play a vital role in managing respiratory disorders such as chronic obstructive pulmonary disease (COPD), asthma, and sleep apnea. Patients with these conditions often experience fluctuating oxygen levels, particularly during physical activity or at night.

Wearable pulse oximeters continuously measure blood oxygen saturation levels (SpO_2), providing real-time data that can be monitored by healthcare providers. In the case of sleep apnea, for example, wearable devices can track changes in oxygen levels during sleep, alerting both the patient and their healthcare provider to any potentially dangerous dips in oxygen saturation that may indicate an apnea event. Similarly, for patients with COPD, wearable devices that measure both oxygen saturation and respiratory rate can provide early warnings of exacerbations, allowing clinicians to intervene with medication adjustments or other treatments before the patient requires hospitalization.

Wearable technology has revolutionized respiratory care in remote, community, and home health care by providing continuous, real-time monitoring of key respiratory parameters, such as blood oxygen saturation (SpO_2), respiratory rate, heart rate variability, and airflow patterns. These devices—ranging from pulse oximeters to smart respiratory wearables—integrate with connected health platforms, allowing healthcare providers to track disease progression, detect early warning signs, and intervene before complications arise.

COPD is a progressive respiratory disease characterized by airflow limitation and breathing difficulties. Patients with COPD often experience fluctuations in oxygen levels, especially during exertion or sleep. Early detection of exacerbations is critical, as timely interventions can reduce hospitalizations and improve quality of life.

- **Continuous Oxygen Saturation Monitoring**: Wearable pulse oximeters continuously measure SpO_2 levels, helping patients and providers identify trends that indicate respiratory decline. A sudden drop in oxygen saturation may signal an impending exacerbation, prompting healthcare teams to initiate treatment adjustments, such as increased bronchodilator use, supplemental oxygen, or corticosteroids.
- **Respiratory Rate and Breathing Pattern Analysis**: Some advanced wearables can monitor respiratory rate and detect changes in breathing patterns, such as shallow breathing or prolonged exhalation, which are indicative of worsening COPD. By analyzing real-time data, AI-driven algorithms can identify anomalies and send alerts to both patients and providers.
- **Activity and Symptom Tracking**: COPD patients are often advised to stay physically active within safe limits. Smart wearables with accelerometers can track

physical activity levels, steps taken, and exertion-related symptoms such as breathlessness. If a patient becomes less active due to worsening symptoms, healthcare providers can intervene promptly to prevent further decline.
- **Telemedicine Integration**: Wearable devices in COPD management can sync with telehealth platforms, allowing patients to consult their healthcare providers remotely. This reduces the need for frequent hospital visits and ensures that care is continuous and accessible.

Asthma Management and Connected Wearables

Asthma is a chronic inflammatory condition of the airways that causes breathing difficulties, wheezing, and coughing. Environmental factors, allergens, exercise, or infections can trigger symptoms. Wearable devices enhance asthma management by enabling continuous monitoring of lung function and environmental exposure.

- **Smart Wearable Spirometers**: Some wearable devices measure lung function parameters such as forced expiratory volume (FEV1) and peak expiratory flow rate (PEFR), which are essential for assessing airway obstruction. These devices can be used at home to track lung function trends, helping patients and clinicians adjust treatment plans accordingly.
- **Respiratory Rate and Wheeze Detection**: Certain smart patches and chest-worn wearables can detect abnormal breathing patterns and wheezing sounds, providing early warnings of asthma exacerbations. If a patient experiences increased airway resistance, the device can trigger an alert, prompting the patient to use rescue medication or seek medical attention.
- **Air Quality Monitoring and Personalized Alerts**: Connected health platforms integrate data from environmental sensors to assess air quality, pollen levels, and pollution indices. Patients with asthma can receive personalized alerts advising them to avoid high-risk environments or adjust their medication use based on real-time conditions.
- **Medication Adherence Tracking**: Smart inhalers, equipped with sensors, track medication usage and inhaler technique. Data from these devices is shared with healthcare providers, ensuring that patients adhere to prescribed treatment plans and use their inhalers correctly.

Sleep Apnea Monitoring with Wearable Devices

Sleep apnea is a disorder characterized by repeated interruptions in breathing during sleep, leading to fragmented sleep patterns, daytime fatigue, and an increased risk of cardiovascular complications. Traditional sleep apnea diagnosis requires in-lab polysomnography, but wearable devices now enable home-based monitoring and early detection.

- **Wearable Pulse Oximeters and Sleep Trackers**: Many wearable devices, including smartwatches and finger-worn pulse oximeters, can monitor SpO_2 levels throughout the night. A pattern of oxygen desaturation, especially in conjunction with changes in heart rate and respiratory rate, may indicate sleep apnea events.

- **Smart CPAP Integration**: For patients using continuous positive airway pressure (CPAP) therapy, wearable sensors can sync with CPAP machines to track usage patterns, mask fit, and respiratory responses. This data is uploaded to connected health platforms, allowing sleep specialists to optimize therapy settings remotely.
- **AI-Powered Sleep Pattern Analysis**: Advanced wearables utilize AI algorithms to detect sleep apnea-related disturbances based on movement, respiratory effort, and oxygen saturation levels. These devices can generate automated reports, aiding in early diagnosis and treatment planning.

The Future of Connected Health in Respiratory Care

The integration of wearable technology with connected health platforms is reshaping respiratory disease management in several ways:

- **Early Detection and Preventive Care**: By continuously tracking vital signs and respiratory parameters, wearable devices can identify disease exacerbations early, allowing for timely interventions and reducing emergency hospitalizations.
- **Personalized Treatment Plans**: AI-powered analytics can assess individual patient data, enabling personalized treatment recommendations tailored to each patient's unique condition and lifestyle.
- **Improved Patient Engagement**: Real-time feedback from wearables empowers patients to take a more active role in managing their condition. By visualizing their health data, patients are more likely to adhere to treatment plans and make informed decisions about their well-being.
- **Remote Healthcare Accessibility**: Connected health solutions reduce the burden on healthcare facilities by enabling remote monitoring and telehealth consultations. This is especially beneficial for patients in rural or underserved areas who may have limited access to specialists.
- **Integration with Smart Home Systems**: Future advancements in connected health will likely see wearables integrating with smart home technologies, such as AI-powered voice assistants and air quality monitors, to provide an even more comprehensive approach to respiratory health management can track physical activity levels, steps taken, and exertion-related symptoms such as breathlessness. If a patient becomes less active due to worsening symptoms, healthcare providers can intervene promptly to prevent further decline.

1.5 Postoperative Monitoring

Connected health is revolutionizing postoperative monitoring by utilizing wearable devices that provide real-time tracking of vital signs, activity levels, and wound healing, significantly reducing the risk of complications and hospital readmissions. Traditionally, post-surgical patients require frequent follow-up visits to assess recovery progress, monitor for infections, and ensure proper healing. However, wearable devices such as smart patches, biosensors, and wireless monitoring systems enable continuous remote surveillance, allowing clinicians to track parameters like heart rate, respiratory rate, temperature, and oxygen saturation.

From the first types of monitoring to include weight, blood pressure and temperature that was entered into a patient note from telecommunications each morning, and per the physician order for frequency, these devices have developed over time to detect early signs of complications, such as infection or blood clot formation, prompting timely medical intervention before the condition worsens. Additionally, smart bandages equipped with sensors can monitor wound healing by analyzing temperature variations and moisture levels, alerting healthcare providers to potential issues like delayed healing or surgical site infections.

Wearable motion trackers further aid in postoperative rehabilitation by assessing mobility and physical activity, ensuring that patients regain function at an appropriate pace. By integrating these wearables with connected health platforms, physicians can make data-driven decisions, optimize pain management, and provide personalized recovery plans without requiring patients to make frequent hospital visits. This seamless connectivity not only improves patient outcomes but also enhances healthcare efficiency by reducing strain on hospital resources and improving the overall recovery experience.

1.6 Mental Health and Well-being

In addition to physical health, connected health is increasingly playing a role in mental health and well-being by leveraging wearable devices to monitor sleep patterns, stress levels, and physical activity. Mental health conditions such as anxiety, depression, and chronic stress are often associated with physiological markers, including heart rate variability, sleep disturbances, and fluctuations in cortisol levels.

The role of wearable technology in mental health management is expanding, with devices capable of tracking physiological and behavioral indicators associated with emotional well-being. Smartwatches and fitness trackers monitor sleep patterns, activity levels, and heart rate variability, providing valuable insights into stress levels and overall mental health. Some wearables also incorporate electrodermal activity (EDA) sensors, which detect changes in skin conductivity related to stress responses.

Advanced wearables now integrate mood-tracking features that analyze voice patterns, facial expressions, and physiological responses to assess emotional states. By integrating with mobile health applications, these devices offer users personalized recommendations for stress reduction, meditation, and lifestyle adjustments. Additionally, wearable technology supports behavioral health interventions by allowing clinicians to monitor patient progress remotely and make data-driven decisions regarding mental health treatment. Emerging applications in mental health wearables also include biofeedback training, which helps users gain better control over physiological responses to stress. As awareness of mental well-being grows, wearable devices continue to play a crucial role in promoting self-care and improving access to mental health resources. These advancements pave the way for more personalized and proactive mental health management, offering real-time support and interventions tailored to individual needs.

By enhancing disease management, enabling early detection, improving postoperative recovery, and supporting mental health, wearable devices are becoming integral components of modern healthcare systems. Their continuous evolution is expected to further transform patient care by making health monitoring more accessible, efficient, and data-driven. As research and development continue to push the boundaries of wearable healthcare technology, future iterations of these devices will likely incorporate even more sophisticated capabilities, such as real-time blood analysis, AI-driven diagnostics, and integration with neural health monitoring. The ongoing advancements in wearable technology are shaping a future where remote patient care is more personalized, predictive, and preventative than ever before.

By providing continuous, non-invasive support, wearable technology empowers users to take proactive control of their mental well-being while enabling healthcare professionals to monitor at-risk individuals and intervene when necessary remotely. The seamless integration of wearables with connected health platforms is transforming mental healthcare by promoting self-awareness, early intervention, and personalized support, ultimately enhancing overall well-being.

1.7 Empowering Patients and Improving Health Outcomes

One of the most transformative aspects of wearable devices in chronic disease management is the empowerment they offer patients. By providing continuous access to health data, wearables enable patients to monitor their conditions in real time and make informed detect early signs of deterioration, and prevent potentially life-threatening events. Wearable ECG (electrocardiogram) devices, such as those integrated into smartwatches, allow patients to track their heart rhythms continuously. These devices can automatically upload data to a healthcare provider's dashboard, enabling real-time analysis by physicians.

For instance, if a patient with known atrial fibrillation experiences irregular heartbeats, their wearable ECG monitor will capture and transmit this data to their healthcare team, triggering an alert. The clinician can then review the data, confirm the diagnosis, and recommend an intervention, such as adjusting medication or scheduling a follow-up appointment. This early detection capability significantly reduces the risk of serious complications, such as stroke or heart failure exacerbations, by ensuring that patients receive timely and appropriate care.

Security and Privacy Considerations

As wearables collect vast amounts of sensitive personal health data, security and privacy are paramount. Secure data transmission protocols, encryption, and robust authentication measures are critical to ensuring that patient data remains protected throughout the exchange process. Healthcare organizations must also comply with regulatory standards, such as HIPAA in the United States, to safeguard patient privacy and ensure that data is used appropriately.

A key advancement in connected health is interoperability, which allows wearable devices to seamlessly integrate with electronic health records (EHRs), cloud-based analytics, and artificial intelligence (AI)-driven diagnostic tools. Wearables collect vast amounts of physiological data—such as heart rate, oxygen saturation, glucose levels, and sleep patterns—and transmit it to secure digital health platforms via wireless technologies like Bluetooth, Wi-Fi, and 5G networks.

This real-time data exchange provides healthcare providers with a continuous stream of patient information, eliminating the need for patients to rely solely on periodic clinical visits for monitoring chronic conditions. For example, a patient with cardiovascular disease can wear an ECG-enabled smartwatch that automatically uploads heart rhythm data to their physician's dashboard, allowing for early detection of abnormalities and timely intervention.

1.8 Enhancing Connected Health

The convergence of wearables, connected health platforms, and AI-driven analytics has unlocked new possibilities in predictive healthcare. By analyzing continuous streams of patient data, AI can detect subtle deviations in physiological patterns and predict potential health risks before symptoms manifest.

For example, AI-enabled wearables can detect irregular heart rhythms that may indicate early signs of atrial fibrillation (AFib), prompting users to seek medical attention before a life-threatening cardiac event occurs. Similarly, smart glucose monitors integrated with AI can identify blood sugar fluctuations and provide real-time dietary and medication recommendations, helping diabetic patients better manage their condition.

Furthermore, connected health platforms leverage machine learning algorithms to personalize healthcare recommendations based on individual health profiles. For example, a wearable fitness tracker can analyze a user's sleep patterns, stress levels, and physical activity trends to offer customized wellness plans aimed at improving overall health and preventing chronic diseases.

Challenges and Considerations in Integration

While the integration of wearable devices with connected health platforms offers numerous benefits, it also presents several challenges that must be addressed for widespread adoption.

Data Privacy and Security

One of the biggest concerns surrounding wearable technology is the security of patient data. With large volumes of health data being transmitted wirelessly and stored in cloud systems, the risk of cyber threats and unauthorized access increases.

Implementing robust encryption methods, blockchain technology, and stringent regulatory policies such as GDPR and HIPAA compliance can help safeguard patient information.

Interoperability Issues

Despite advancements in digital health infrastructure, interoperability remains a major hurdle. Different wearable device manufacturers use proprietary software and data formats, making it difficult for healthcare providers to integrate data from various sources. Standardizing data formats and creating universal APIs (Application Programming Interfaces) can enhance seamless connectivity across different healthcare systems.

Regulatory and Ethical Concerns

The regulatory landscape for wearable medical devices is still evolving. Ensuring that wearables meet medical-grade accuracy and are approved by organizations such as the FDA and EMA is crucial for their integration into healthcare. Ethical concerns also arise regarding continuous monitoring and potential misuse of health data by third parties, necessitating strict governance frameworks.

User Adoption and Engagement

For wearable devices to be effective in remote healthcare, patient engagement is essential. Many users abandon their devices due to discomfort, complexity, or lack of perceived value. Designing user-friendly wearables with intuitive interfaces, ergonomic designs, and real-time feedback mechanisms can improve long-term adherence.

1.9 Conclusion: The Future of Wearable Devices in Connected Health

The integration of wearable devices in connected health has revolutionized the landscape of remote patient care, bridging the gap between traditional healthcare settings and continuous, real-time monitoring. These technologies provide a seamless blend of convenience, personalization, and efficiency, allowing patients to actively participate in their healthcare while giving providers access to continuous streams of critical health data. As the healthcare industry shifts towards preventive and personalized medicine, wearable devices are becoming indispensable tools in the management of chronic diseases, postoperative recovery, mental health, and overall well-being.

One of the most transformative aspects of wearable health technology is its ability to foster patient engagement. By providing individuals with real-time insights into their heart rate, glucose levels, oxygen saturation, sleep patterns, and activity levels, wearables empower users to take proactive steps toward improving their health. This increased awareness can lead to better adherence to treatment plans, lifestyle modifications, and early interventions, ultimately reducing the burden of chronic conditions and hospitalizations.

From a clinical perspective, wearables enhance medical decision-making by providing physicians with longitudinal health data that extends beyond isolated clinical visits. This continuous monitoring allows for early detection of potential health issues, such as cardiac arrhythmias, respiratory decline, or glucose fluctuations, enabling timely interventions that prevent complications. Additionally, AI-driven analytics integrated into wearable platforms further enhance diagnostic accuracy and treatment recommendations, making healthcare more data-driven and personalized.

Despite these advancements, challenges remain in the widespread adoption and integration of wearables into mainstream healthcare systems. Issues related to data security, interoperability, regulatory compliance, and patient privacy must be addressed to ensure that wearable-generated health data can be securely shared across electronic health record (EHR) systems, telehealth platforms, and clinical decision support tools. Moreover, health disparities and accessibility concerns need to be considered, ensuring that all patients, regardless of socioeconomic status, can benefit from the latest advancements in wearable health technology.

However, ongoing technological advancements, improved regulatory frameworks, and evolving healthcare policies are paving the way for a future where wearables become seamlessly embedded into routine healthcare practices. The integration of 5G, artificial intelligence, blockchain for data security, and cloud-based analytics will further strengthen the role of wearables in remote monitoring, early diagnosis, and population health management. As healthcare systems continue to prioritize value-based care models focused on cost reduction and improved patient outcomes, wearables will play an increasingly central role in reducing hospital readmissions, optimizing chronic disease management, and enhancing preventive care strategies.

Ultimately, wearable devices are not just passive health-tracking tools—they are reshaping the very foundation of modern healthcare. By enabling continuous, real-time, and personalized monitoring, fostering proactive patient engagement, and driving clinical efficiency, these technologies represent a paradigm shift in how healthcare is delivered. As innovation continues to drive the evolution of connected health, wearables will undoubtedly remain at the forefront, ushering in a new era of intelligent, data-driven, and patient-centered healthcare.

As wearable technology continues to evolve, its integration with connected health will become even more sophisticated. Emerging innovations such as implantable biosensors, remote robotic-assisted diagnostics, and blockchain-secured health data exchanges will further enhance the security, accuracy, and efficiency of connected healthcare systems.

Additionally, regulatory advancements and industry-wide standardization will drive greater adoption of wearable-integrated healthcare solutions, ensuring that these technologies remain scalable, accessible, and seamlessly interoperable across various medical institutions and healthcare networks.

Ultimately, the integration of wearables with connected health platforms is transforming them from standalone consumer gadgets into indispensable tools for preventive care, chronic disease management, and personalized medicine. This convergence is ushering in a new era of digital healthcare, where real-time data, remote monitoring, and AI-driven insights empower both patients and healthcare providers to achieve better health outcomes with greater efficiency and convenience.

References

1. Awad, A., Trenfield, S.J., Pollard, T.D., Ong, J.J., Elbadawi, M., McCoubrey, L.E., Goyanes, A., Gaisford, S., Basit, A.W.: Connected healthcare: improving patient care using digital health technologies. Adv. Drug Deliv. Rev. **178**, 113958 (2021)
2. Cerrato, P., Halamka, J.: Realizing the promise of precision medicine: the role of patient data, mobile technology, and consumer engagement, pp. 101–107. Academic Press (2017)
3. Chodankar, D., Raval, T.K., Jeyaraj, J.: The role of remote data capture, wearables, and digital biomarkers in decentralized clinical trials. Perspect. Clin. Res. **15**(1), 38–41 (2024)
4. Jean-Louis, G., Seixas, A.A.: The value of decentralized clinical trials: Inclusion, accessibility, and innovation. Science **385**(6711), eadq4994 (2024)
5. Sehrawat, O., Noseworthy, P.A., Siontis, K.C., Haddad, T.C., Halamka, J.D., Liu, H.: Data-driven and technology-enabled trial innovations toward decentralization of clinical trials: opportunities and considerations. Mayo Clinic Proc. **98**(9), 1404–1421 (2023)
6. Shafik, W.: Artificial intelligence-Enabled Internet of Medical Things for enhanced healthcare systems. In: Smart Healthcare Systems, pp. 119–134. CRC Press, Boca Raton (2024a)
7. Shafik, W.: Artificial intelligence-enabled Internet of Medical Things (AIoMT) in modern healthcare practices. In: Clinical Practice and Unmet Challenges in AI-Enhanced Healthcare Systems, , pp. 42–69. IGI Global (2024b)
8. Shafik, W.: Connected healthcare—the impact of Internet of Things on medical services. In: Artificial Intelligence and Internet of Things based Augmented Trends for Data Driven Systems, pp. 181–217. CRC Press, Boca Raton (2024c)
9. Shafik, W.: Dissecting the role of women in cybersecurity and information technology: a medical perspective. In: Next-Generation Cybersecurity: AI, ML, and Blockchain, pp. 325–350. Springer (2024d)
10. Shafik, W.: Smart health revolution: exploring artificial intelligence of internet of medical things. In: Healthcare Industry Assessment: Analyzing Risks, Security, and Reliability, pp. 201–229, Springer (2024e)
11. Shafik, W.: Wearable medical electronics in artificial intelligence of medical things. In: Handbook of Security and Privacy of AI-Enabled Healthcare Systems and Internet of Medical Things, pp. 21–40 (2024f)
12. Zundel, K.M.: Telemedicine: history, applications, and impact on librarianship. Bull. Med. Libr. Assoc. **84**(1), 71 (1996)
13. Fye, W.B.: A history of the origin, evolution, and impact of electrocardiography. Am. J. Cardiol. **73**(13), 937–949 (1994)
14. Caddick, B.: Remote consultations: a prescription from the past? Prescriber **34**(4), 19–22 (2023)
15. Blank, E., Lappan, C., Belmont, P.J., Jr., Shaun Machen, M., Ficke, J., Pope, R., Owens, B.D.: Early analysis of the United States Army's telemedicine orthopaedic consultation program. J. Surg. Orthop. Adv. **20**(1), 50 (2011)
16. Singh, S., Kumar, P., Rehman, F., Vashishta, P.: Telemedicine, telehealth, and e-health: a digital transfiguration of standard healthcare system. In: Cloud IoT, pp. 143–161. Chapman and Hall/CRC (2022)
17. Pool, S.L., Belasco, N.: An integrated medical system for long-duration space missions. J. Spacecr. Rocket. **9**(8), 613–614 (1972)
18. Meyers, L., Gibbs, D., Thacker, M., LaFile, L.: Building a telehealth network through collaboration: the story of the Nebraska Statewide Telehealth Network. Crit. Care Nurs. Q. **35**(4), 346–352 (2012)

19. Laouyane, A.: Telemedicine and developing countries. J. Telemed. Telecare **4**(2_suppl), 1–88 (1998)
20. Tricoli, A., Nasiri, N., De, S.: Wearable and miniaturized sensor technologies for personalized and preventive medicine. Adv. Func. Mater. **27**(15), 1605271 (2017)
21. Devi, D.H., Duraisamy, K., Armghan, A., Alsharari, M., Aliqab, K., Sorathiya, V., Rashid, N.: 5g technology in healthcare and wearable devices: a review. Sensors **23**(5), 2519 (2023).
22. Sadrawi, M., Lin, C.H., Lin, Y.T., Hsieh, Y., Kuo, C.C., Chien, J.C., Shieh, J.S.: Arrhythmia evaluation in wearable ECG devices. Sensors **17**(11), 2445 (2017)
23. Baliga, S., Basarakod, R., Kalathmika, G., Pillai, N.P., Shivakumar, J., Yadav, P.: Applications of machine learning to the impact of IoT in biomedical applications. In: Internet of Things enabled machine learning for biomedical applications, pp. 77–96. CRC Press (2024)
24. Adepoju, V.A., Jamil, S., Biswas, M.S., Chowdhury, A.A.: Wearable technology in the management of chronic diseases: a growing concern. Chronic Dis. Transl. Med. (2024)
25. Chiauzzi, E., Rodarte, C., DasMahapatra, P.: Patient-centered activity monitoring in the self-management of chronic health conditions. BMC Med. **13**, 1–6 (2015)

Dr. Wasswa Shafik is a prominent researcher and IEEE member currently leading the Dig Connectivity Research Laboratory (DCRLab) in Kampala, Uganda. He holds a Ph.D. in Computer Science from the School of Digital Science, Universiti Brunei Darussalam. He also possesses a Master's degree in Information Technology Engineering from Yazd University, Iran, and a Bachelor's degree from Ndejje University, Uganda. Dr. Shafik's diverse experience includes roles at NIH, US Department of Health and Human Services, and the Bloomberg School of Public Health. His research is centered around artificial intelligence, computer vision, neural networks, and the Internet of Things, particularly applied in smart agriculture, digital health, and ecological informatics.

Decoding the Potentiality of Wearable Devices in Healthcare: A Mixed Methodology Approach

Hemant Gupta, Bhavesh Kumar J. Parmar, and Kalgi Shah

Abstract Across the globe, the habitude for wearable devices (WD) has come in handy for the routine checking and monitoring of health-related matters in recent times. However, challenges like adaptation, implementation, and others have caused hindrances in the optimum utilization of the product. This research has tried to open up various windows for the usage of WD in the healthcare domain and the challenges followed by the adaptation of the technology. It is a study focusing on customer survey-based insights on comfort, feedback, and the analysis of WD in healthcare along with the use of data analysis to understand the adaptation of it. It studies the Technology Acceptance Model (TAM), thus, bringing up its relevance on its perceived usefulness and its ease of use. The results of the customer survey-based insights on the adoption of WD. It also brings into focus, factors like the level of comfort associated with the use of WD and factors affect the adaptation of it. This study has tried to bring up the challenges in the adaptation of WD in healthcare, its potential in marketing research, and implementation and factors that would be of sheer help in bringing out the optimum utilization of WD.

Keywords Wearable devices · Healthcare · Technology adoption · Technology acceptance model · Mixed method approach

1 Introduction

The integration of wearable devices (WD) in healthcare has yielded revolutionary results in terms of patient outcomes and cost monitoring, because of technological advancements. The exciting thing about these WD is that they can track an array of indicators, such as blood pressure, heart rate, and level of activity, giving

H. Gupta (✉)
ICFAI Foundation for Higher Education, Bangalore, India
e-mail: hemant.bpm13@gmail.com

B. K. J. Parmar · K. Shah
School of Management Studies, National Forensic Sciences University, Gandhinagar, India

P. Eappen et al. (eds.), *Remote Monitoring and Wearable Devices in Healthcare*, Information Systems Engineering and Management 63,
https://doi.org/10.1007/978-3-031-98897-4_3

users comprehensive information about their health [1]. Moreover, it is a monitoring system, it assists patients in leading balanced lives, improving their ability to control their health, and preventing chronic illness. Smartwatches, fitness trackers, continuous glucose monitoring (CGM), smart clothes, smart technologies, and other devices that can track and send health data are examples of WD in healthcare industry. These devices have biosensor technology built into them that can monitor a variety of health parameters, such as blood flow, heart rate, sleep patterns, and others. This technology helps medical practitioners provide patients with, treatments programs that are advanced in advance [2].

Regardless of using WD in healthcare. Due to a lack of standardization, it encounters numerous difficulties while integrating multiple devices, making it challenging for professionals to practice. Patients' confidence in technology may be harmed by privacy and security issues, which could pose a threat in the event of hacking or data breaches [3]. Furthermore, patients' skepticism about the efficacy of WD could leave patients and healthcare providers at a disadvantage. In addition to disease prevention and monitoring, WD in the healthcare industry face opportunities and problems related to marketing. Providing a hassle free and customized experience is one of the primary marketing chances. Utilizing the information gathered about consumer behaviour and preferences, healthcare providers can provide more individualized and efficient care [4]. WD also faces difficulties in protecting the security and privacy of its data. Another factor is that even knowing about the dangers to security, privacy, usability, etc., users continue to use WD. A major concern for both patients and medical professionals is the high cost of WD [5].

Despite all of these barriers, the WD industry has grown significantly, and in the upcoming years, technology businesses as well as healthcare providers anticipate seeing a sharp increase. The development of artificial intelligence (AI), has improved WD and increased its benefits for the healthcare industry. An in-depth examination of the potential advantages and barriers linked to WD in the healthcare sector can assist healthcare providers and technology firms in making well-informed choices regarding investment in this technology and subsequently shaping their marketing strategy.

2 Research Problems and Objectives

The goal of this research is to carry out a thorough marketing analysis while taking into account WD's potential in healthcare, notwithstanding the difficulties this industry faces. This study aims to investigate the marketing potential of wearable devices (WD) in the healthcare industry, as well as the opportunities and problems that arise for marketers in this regard as WD adapts. The data study will highlight WD's advantages and disadvantages, point out market prospects and challenges, and assess the state of wearable technology in the healthcare sector at present. Furthermore, the customer- based survey will help gather input from medical professionals who actually use these devices in practice as well as offer insightful information on how patients feel about the comfort and usefulness of WD. Finally, a framework outlining the factors influencing patients and healthcare professionals' adaption and acceptance of WD will be made available by the TAM evaluation.

The prospective advantages of utilizing WD in healthcare and the requirement for marketing professionals to comprehend the prospects and obstacles related to its adaption are what provide this study its special significance. With its ability to track blood pressure, heart rate, glucose levels, and other vital signs. WD's usefulness in the healthcare industry is booming for both patients and professionals. Thus, facilitating improved self-care and chronic illness prevention. By monitoring the health data and raising patient satisfaction, it also helps to reduce the cost of hospital admissions by educating patients in advance about their illnesses.

3 Literature Review

The WD's capacity to deliver real time patient health status data, which offers valuable insights to medical practitioners, has helped it gain global recognition in recent years. Smartwatches, fitness trackers, and medical devices are only a handful of many health-care related products that WD offers functionality for that can monitor and transfer health data. These electronic devices have Nano sensors built in that can monitor a user's blood pressure, heart rate sleep schedule, and other parameters. Healthcare practitioners can monitor patient's health, identify problems, and create individualized treatment plans with the help of this data [3].

By enhancing patient outcomes and cutting costs, WD offer several advantages in the healthcare industry. By providing access to real-time patient health status data, it improves patient outcomes by assisting healthcare providers in identifying and addressing healthcare issues. Additionally, patients benefit from it since it helps them maintain healthier living habits, practice self- management, and avoid chronic illnesses by keeping track of their medical information [6]. Moreover, by offering more precise and impartial data on patient's health status, WD have the potential to completely transform clinical trials. Instead than depending mostly on patient self- reports, as is commonly the case with traditional clinical trial approaches. WD can provide more thorough and accurate information by gathering data on patient's health state instantly. Notwithstanding all the previously listed advantages, there are certain drawbacks that affect how quickly technology is adapted. Concerns about patient data security and privacy rank among the biggest obstacles. These devices have the ability to track, gather and send patient health data, which increases the risk of data compromise and possible patient harm. Therefore, guidelines and standards are required to guarantee the precision and dependability of WD in the medical field [7].

WD, often referred to as wearables, are electronic devices that are worn on the body that have the ability to continually measure and monitor parameters relevant to one's health [8]. Vital sign data, such as heart rate, blood pressure, breathing rate, sleep patterns, etc. Can be gathered using it. Additionally, it focuses on the observation and treatment of long-term conditions like diabetes, heart disease, and hypertension.

3.1 Types of WD in Healthcare

There are four different kinds of wearable devices (WD): body sensors, wristbands, smart apparel, and smart watches. The most popular ones are wristbands and smartwatches, which are mostly intended to track user's everyday activities [9]. As opposed to this, body sensors are affixed to the body in order to track numerous metrics such as blood pressure, glucose levels, and other vital indications. An emerging form of wearable technology called "smart clothing" uses sensors and electronics to monitor physiological characteristics including body temperature, heart rate, and respiration rate [10].

3.2 Opportunities and Challenges of WD

WD may act as an innovative tool which reduces the healthcare cost by its constant monitoring and early diagnosis. While it becomes extremely useful in the treatment of chronic illnesses, its main usefulness is the accumulation of vast amounts of real–time data. WD is capable of tracking critical indicators of patient's health, such as blood pressure, oxygen saturation, heart rate and glucose flow. In order for medical practitioners to create individualized treatment programs based on each patient's needs, this information is crucial [11].

The development of telemedicine by WD has also had a significant impact. The practice of providing medical care remotely using technology is known as telemedicine. Frequent trips to medical facilities can be avoided by using WD to remotely monitor patients. Healthcare professionals can now reach patients who live in rural places with limited access to medical care because to telemedicine. Telemedicine allows patients to get healthcare from the comfort of their homes, which can be advantageous for people with restricted mobility or those taking frequent medications [12]. The potential in WD is enormous. When it comes to managing chronic illnesses, it can be of great assistance. In international marketplaces, chronic illnesses are a verified source of expense. WD may be able to offer early diagnosis, which would allow for early action and a decrease in expenses. Through real- time monitoring of drug intake and physical activity, it can also decrease hospital admissions and increase treatment adherence [13].

There are numerous challenges that stand in the way of the technology adoption and application, despite its potential in the larger market. Its privacy and security issues rank among the most significant issues. A lack of effective security could lead to hacking or unauthorized access to the vast amounts of personal data that WD collects. Patients and healthcare professional's adoption of WD may be questionable because to the absence of established regulations. To guarantee the security and effectiveness of WD, a thorough regulatory framework is required [14].

3.3 *Technology Acceptance Model and Its Relevance to WD in Healthcare*

One theoretical framework that aids in forecasting the adoption of new technologies is the Technology Acceptance Model (TAM). It explains how the perceived utility and usability of new technology affects adoption. Over the years, TAM has been used to evaluate novel technology, like the WD in this instance. WD, especially those used in healthcare environments, necessitate sophisticated communication between the patient, the healthcare professional, and technology. Therefore, patient and healthcare provider acceptance and implementation of the technology is critical to the success of WD in healthcare. For this reason, the TAM model is essential for comprehending the use of WD in healthcare.

3.4 *Perceived Usefulness*

In TAM, perceived usefulness is one of the factors that affects how technology is adopted. Perceived usefulness, as defined by TAM, is the extent to which a person feels that a technology can enhance their quality of life or performance at work. In the context of medicine, it refers to the extent to which practitioners and patients alike think that WD may raise standards of health and lower medical expenses. Individuals can receive continuing health progress reports and real- time health data from WD. With the help of information supplied by WD, patients may make better decisions about their health and take action to enhance their results [15].

3.5 *Perceived Ease of Use*

Perceived ease of use is an another significant determinant of TAM. It refers to the extent to which a person thinks using technology is simple and convenient. It should be easy for both users to traverse WD and should be user- friendly. The adoption of WD in healthcare may be discouraged if they are difficult to operate. Additionally, healthcare professionals must receive training on how to analyse and incorporate WD's data in to their clinical procedures [16]

3.6 Comfort

For WD to be successful in the healthcare industry, comfort is essential. It must be comfortable for patients to wear these devices for prolonged periods of time. The [17] assessed 200 individuals who had sleep apnea and were wearing a WD. On a scale of 1 to 10, the participants were asked to rate how comfortable they were. The average degree of comfort, according to the findings, was 7.8 out of 10.

3.7 Feedback

Another crucial element in WD's success in healthcare is feedback. In order for patients to comprehend their current state of health and make wise decisions, they must obtain feedback on their health data. The [18] polled fifty patients who used a WD for continuous glucose monitoring as part of their study. The participants were questioned regarding the device's use and the comments they had received. According to the findings 86% of the participants thought the feedback helped them manage their diabetes, and 94% of participants thought the gadget was useful.

3.8 Analysis

In order to supply medical expert insightful information about a patient's health, WD's data analysis is essential. The [19] examined the information gathered by WD in order to find trends in the health and behaviour of her patients. The information demonstrated that WD could provide useful information about patient's heart rates, sleeping patterns, and physical activity.

In recent years, WD has become more widely used in the healthcare industry. It has been discovered that WD are useful for tracking a variety of health indicators and enhancing patient outcomes. Research utilizing secondary data analysis has demonstrated the applicability of WD in a range of healthcare contexts, including elder care, cancer survivorship, and rehabilitation. Healthcare institutions can save money and boost patient motivation and involvement by implementing WD. But there are still issues that need to be resolved regarding WD's accuracy, dependability, and privacy. Further studies ought to concentrate on resolving the constraints of WD in the medical field, like enhancing their precision and dependability. To enhance patient outcomes and save money, it is critical that healthcare providers accept the use of WD and integrate it in to their practices.

4 Research Methodology

4.1 Research Design

The following are the study questions: what are the marketing opportunities and obstacles that prevent WD from being used in healthcare? It is hypothesized that while adapting WD can result in substantial marketing potential, there are certain challenges that must be overcome for the implementation to be successful. This study's main objective is to identify the possible advantages and challenges related to WD adaption in the healthcare industry. The primary objective of this research is to ascertain the elements that facilitate the effective integration of WD and the obstacles that must be overcome to guarantee their successful execution.

The study will use a mixed method analysis, which will include a thorough secondary data analysis, consumer survey-based insights on comfort, feedback, analysis, and TAM evaluation, to test the hypothesis. The secondary data analysis sheds light on the current situation of the WD market and offer insights in to the trends, challenges, and opportunities associated with its adaption. The examination of the customer survey will shed significant light on the preferences, attitudes and perceptions of the customers regarding WD in healthcare. The TAM review will provide light on the variables affecting customer's adoption and acceptability. According to the study hypothesis, enhanced patient involvement, lower healthcare costs, and better patient outcomes are among the marketing potential connected to the adoption of WD in the healthcare industry. Reluctance to change, inadequate interoperability, and concerns about data security and privacy are some of the marketing challenges that come with WD adoption in the healthcare industry.

This study is structured using a mixed methods approach which incorporates both qualitative and quantitative research techniques. A survey-based study is part of this undertakings initial phase, which collects information on how customers view WD. To learn more about the usage habits, satisfaction levels, and preferences of a sample of healthcare customers, a structured questionnaire is created and given to them. The TAM is presented in the second stage to assess WD consumers' acceptance in the healthcare system.

4.2 Sampling Techniques and Sampling Size

A research study's sample size and sampling strategy are essential elements. Purposive sampling will be utilized in this study to choose participants who fit the inclusion requirements. Purposeful sampling is a non-probability selection strategy that allows study participants to be selected based on specific characteristics that make them eligible for this research.

The participants who have utilized WD for healthcare and are at least 18 years old will meet the study's inclusion requirements. Based on recommendations from the literature on sample size determination for marketing research, the sample size for this study will be established. The formula for determining sample size for a given population size, as proposed by [20], will be used to determine the sample size. For this study, a minimum of 284nparticipants will be chosen as the sample size. The sample size will be adequate to minimize the margin of error and produce a statistically meaningful result. Hospitals, clinics and pharmacies are just a few of the healthcare facilities from which study participants will be chosen. Online health forums and social media platforms will also be used to attract participants. Prior to participation, participants will give informed consent and buyer fully informed about the study's purpose and objectives.

4.3 Data Analysis Techniques

In this study, we will be using a range of data analysis techniques to analyse the collected data. The first step will be to perform descriptive statistics to summarize and describe the data collected from the secondary sources and the customer survey (see Table 1). Descriptive statistics will be used to provide information about the central tendency, variability, and distribution of the collected data. It will help us to better understand the characteristics of the data and identify any potential outliners or anomalies.

4.4 Results and Findings

Statistics and survey -based insights on comfort, feedback, and analysis of WD in healthcare. The study comprised 284 participants who responded to questions about their age, gender, marital status, level of education, employment status, annual household income, familiarity with and use of wearable devices (WD) in healthcare, reasons for using WD, frequency of use, preferred type of WD, features deemed important in a WD, marketing challenges encountered in adopting WD, factors influencing purchase decisions, perceptions of the usefulness and attitudes towards the use of WD, concerns, improvements in health or wellbeing, challenges faced when using a WD, importance of aesthetics, willingness to pay extra for extra health monitoring features, significance of battery life, are concerned about the privacy and security of health data, likelihood of recommending WD, and importance of integration with other healthcare technologies. The vast majority of respondents (190 out of 300) were in the 18 to 34 age range, with fewer responses in other age groups. Among the participants, 134 identified as female and 150 as male. When asked about their gender, none of the interviewees indicated they were non- binary or that they would remain silent (see Table 1).

Table 1 Demographic description of sample survey here

Age	Under 18	18–35	36–65	Above 65	Total
	10	190	70	14	284
Gender	Male	Female	Non-binary	Prefer not to say	
	150	134	0	0	284
Marital status	Single	Married	Other		
	90	170	24		284
Educational qualification	School level	Under graduate	Post graduate	Doctorate	
	30	150	84	20	284
Annual salaries	Less than 2,00,000 INR	2,00,000–5,00,000 INR	500,000–1,000,000 INR	More than 1,000,000 INR	
	80	120	54	30	284
Familiarity with wearable devices	Very familiar	Somewhat familiar	Not familiar		
	190	89	5		284
Used the wearable device	Yes	No	Prefer not to say		
	220	60	4		284

Source Authors' calculation

According to studies, around 40% of Americans own a fitness tracker or wristwatch, and the volume of data the devices are collecting is increasing. Although wearables are being gradually approved by the FDA data collected by smartwatches and fitness trackers is usually not governed by the same regulations as medical devices, which are subject to stringent regulatory oversight and clinical trials to guarantee data quality, validity, and compliance. All of this can contribute to a rise in customer trust when they are urged to exchange personal information for wellness. Consumer wearables are likely to face skepticism from both users and medical professionals until manufacturers can guarantee data quality and user privacy (see the Table 2).

Apple dominated the market for WD shipments in the fourth quarter of 2023, with a 33.6 percent market share. The wearables industry has a flourishing smartwatch market, with shipments of these devices increasing globally in recent times. Apple leads the world's smartwatch market, accounting for 52.5 percent of shipments worldwide in Q2 2021. Samsung is in second with 11%. When the Apple watch was introduced in 2015, it completely changed the market. In the year following its debut, Apple held a 75% share of the smartwatch market; however, as competition has increased, this share has decreased. With the 2020 release of the Galaxy Watch series, Samsung- which had previously dominated the market- has increased its market share worldwide.

Table 2 Data analysis report on respondents here

Primary reason for usage				
Fitness tracking	Disease management	General health monitoring	Other	
66%	0	30%	4%	
Preferred type				
Smartwatch	Fitness tracker	Smart clothing	Other	
60%	35%	5%	0	
Features considered most important				
Accuracy Of Data	Battery Life	Design And Appearance	Price	Other
45%	20%	20%	10%	5%
Factors influencing the decision to purchase				
Price	Brand reputation	Features	Recommendation from healthcare providers	Other
35%	30%	30%	5%	0

Source Authors' calculation

The wearables market is now dominated by Samsung, Huawei, Xiaomi, and Apple; in 2023, Apple possess a 29.7% share, making it the leading wearables vendor. It also discusses how the growth of the WD market is being fueled by wearables and smartwatches. Wearables are growing substantially faster than smartwatches because of high levels of product innovation, such as the incorporation of premium microphones and speakers in the devices that allow users to interact with virtual assistants like Apple Siri and the Google Assistant. The study also highlights the growing usage of wearables in medicine, with more medical wearable sensors and equipment expected to be shipped in the coming years.

4.5 Implications of the Result

The survey's findings, which focus on how WD is used and perceived in the healthcare industry, have a number of ramifications for stakeholders in the industry, including manufacturers and policy officials. The poll sheds light on the characteristics of people who use WD for healthcare, their main motivations for using these devices and the difficulties and worries that come with using them:

People utilize WD in healthcare for fitness tracking, which is one of the fundamental- or primary- findings of this study, with general health monitoring serving as the secondary finding. Therefore, it may be concluded that the primary use of WD is in prevention as opposed to the treatment of chronic illnesses. When marketing and designing wearable products, healthcare experts and marketers should consider specific criteria, such as highlighting the preventive benefits and making the product accessible to a wider user base. As a result, the product becomes more successful. The product's accuracy is another consideration. Customers are more likely to be dependable when using these WD in the healthcare industry if the accuracy level is higher. Manufacturers ought to concentrate on giving clients precise data and unambiguous instructions on how to interpret the information gathered. Customers will have more faith in the product as a result.

Several respondents have brought attention to the device's battery life in the findings. Therefore, in order to improve the usability and convenience of the devices for the users, the findings also recommend that manufacturers concentrate on the battery life of the devices. Respondents discussed privacy and security concepts as well as ways to affect WD usage. It is the responsibility of manufacturers to ensure that user consent is obtained and that data is securely stored. The difficulties WD presents in the healthcare system are also mentioned in the poll. The majority of responders were mostly concerned about technical issues and data accuracy. In addition to making sure that users receive the necessary training, healthcare practitioners and device manufacturers should offer clear instructions on how to operate the devices. They have to concentrate on enhancing the precision of the information furnished by the devices in order to augment its usefulness and efficiency. Security and privacy issues had been noted in the results. Regulators should create policies and procedures to safeguard user's privacy. These could include policies about the sharing and security of data as well as the use of user information for marketing and research.

The study also sheds light on how individuals feel about WD in the medical field. And the majority of them were helpful in their intended usage and had a positive attitude. They might even suggest the WD to others in need of medical attention, and it could grow to be a valuable resource for advancing public health. The results of the study also revealed that respondents had concerns over the product's appearance. Therefore, design should be the primary emphasis of manufacturing in order to ensure that the product is both useful and aesthetically beautiful.

Furthermore, a sizable portion of respondents indicated that they would be willing to pay extra for further health monitoring services. This implies that there is a market for more complex and sophisticated WD and that producers have to think about funding the creation of these devices. The study offers information on the demographics of people who receive healthcare from WD as well. Most of the responders have a bachelor's degree and are primarily between the ages of 18 and 35. This suggests that younger and better educated customers may find WD for healthcare to be very popular. Before promoting and creating wearables, healthcare professionals and manufacturers should take these factors in to account.

5 Practical Implications

Given that majority of survey respondents were between the ages of 18 and 35, it is possible that WD makers view this age range as their main target market. According to the research, fitness tracking was the primary applications of WD in healthcare, indicating that device manufacturers should prioritize this feature. The fact that a smartwatch was the most popular wearable device (WD) for healthcare indicated that manufacturers should focus on developing smartwatches with healthcare- related features. The need of precise and reliable data collection in wearable healthcare devices was emphasized by the majority of respondents, who considered it to be the most component of WD. Additionally, the security and privacy of the health information that WD gathered also caused concern for the responders. This suggests that in order to ensure user data security and privacy, manufacturers must take action. The results of the poll demonstrate that customers regard aesthetics highly, as many said they are concerned about how the device looks. Accordingly, designers should focus on producing devices that are both aesthetically pleasing and useful.

The results of the survey provide important and noteworthy information about the beliefs, behaviors, and preferences of thirty respondents with regard to the use of WD in the healthcare sector. Married man in the 18–35 age range with a bachelor's degree or above made up the majority of the responders. The majority of the respondents, whose income ranged from 300,000 to 500,000 Indian rupees, had varying income levels. Tamil was the most often ethnic group; while the majority of respondents identified as professionals. Respondents were overwhelmingly very familiar with WD in healthcare; 18% reported using one for medical purposes. With daily use being the most common frequency of use, wearables were primarily utilized for fitness tracking. Wearable technology (WD) had the greatest traction with fitness trackers and smartwatches. A wearable medical devices accuracy of data was its most important feature; its appearance and battery life came in second and third. Several respondents voiced concerns about security and privacy in addition to marketing challenges. But, most respondents said their health or well-being had improved since using for medical purposes. The difficulties with utilization included technical problems and the veracity of the health data that WD provided. A large number of respondents placed a high priority on aesthetics and were willing to pay more for features that included health monitoring. There was a long battery life. In a WD, a length battery life was also considered crucial.

Despite it, the majority of respondents were quite eager to recommend a WD for medical use, and they believed that it was essential that it work in tandem with current medical technology. These findings suggest that WD may be a useful tool for managing healthcare, but in order to ensure widespread adoption and use, privacy and security concerns, accuracy issues, and technological difficulties must be fixed.

6 Limitations of the Study

Despite all the positive ramifications, there are several shortcomings with this research study. The source of the findings is limited due to the tiny sample size. Despite efforts to include a varied sample size, the study only comprised 280 respondents; hence, generalizability of the findings to a larger or other demographic group may be limited. The study's data were only gathered over a brief two- week timeframe. In the long term, a longer time period would have been beneficial to have a complete grasp of client's preferences.

References

1. Greiwe, J., Nyenhuis, S.M.: Wearable technology and how this can be implemented into clinical practice. Curr. Allergy Asthma Rep. **20**, 1 (2020)
2. Wu, M., Luo, J.: Wearable technology applications in healthcare: a literature review. Online J. Nurs. Inform. **23**(3) (2019)
3. Clark, R., Williams, C., Gilchrist, I.D.: Oculomotor control in children with Special Educational Needs (SEN): the development and piloting of a novel app-based therapeutic intervention. Heal. Technol. **11**(4), 919–928 (2021)
4. Firth, J., Torous, J., Nicholas, J., Carney, R., Pratap, A., Rosenbaum, S., Sarris, J.: The efficacy of smartphone-based mental health interventions for depressive symptoms: a meta-analysis of randomized controlled trials. World Psychiatry **16**(3), 287–298 (2017)
5. Topol, E.J.: High-performance medicine: the convergence of human and artificial intelligence. Nat. Med. **25**(1), 44–56 (2019)
6. Kumar, S., Nilsen, W.J., Abernethy, A., Atienza, A., Patrick, K., Pavel, M., Riley, W.T., Shar, A., Spring, B., Spruijt-Metz, D., Hedeker, D.: Mobile health technology evaluation: the mHealth evidence workshop. Am. J. Prev. Med. **45**(2), 228–236 (2013)
7. Chous, A.P.: Pros and cons of using an AI-based diagnosis for diabetic retinopathy. Optometry times. **10**(8), 1 (2018)
8. Lu, L., Zhang, J., Xie, Y., Gao, F., Xu, S., Wu, X., Ye, Z.: Wearable health devices in health care: narrative systematic review. JMIR Mhealth Uhealth **8**(11), e18907 (2020)
9. Klonoff, A.N., Lee, W.A., Xu, N.Y., Nguyen, K.T., DuBord, A., Kerr, D.: Six digital health technologies that will transform diabetes. J. Diabetes Sci. Technol. **17**(1), 239–249 (2023)
10. Ferreira, P.C., Ataide, V.N., Chagas, C.L., Angnes, L., Coltro, W.K., Paixão, T.R., de Araujo, W.R.: Wearable electrochemical sensors for forensic and clinical applications. TrAC, Trends Anal. Chem. **1**(119), 115622 (2019)
11. Elnaggar, A., von Oppenfeld, J., Whooley, M.A., Merek, S., Park, L.G.: Applying mobile technology to sustain physical activity after completion of cardiac rehabilitation: acceptability study. JMIR Hum. Factors **8**(3), e25356 (2021)
12. Sqalli, M.T., Al-Thani, D.: Evolution of wearable devices in health coaching: challenges and opportunities. Front. Digit. Health. **2**(2), 545646 (2020)
13. Channa, A., Popescu, N., Skibinska, J., Burget, R.: The rise of wearable devices during the COVID-19 pandemic: a systematic review. Sensors. **21**(17), 5787 (2021)
14. Franco, P., Martínez, J.M., Kim, Y.C., Ahmed, M.A.: A cyber-physical approach for residential energy management: current state and future directions. Sustainability **14**(8), 4639 (2022)
15. Gupta, H., Singhe, A.: Prediction and analysis of smartphone among customers. In: 2023 International Conference on New Frontiers in Communication, Automation, Management and Security (ICCAMS), Vol. 1, pp. 1–9. IEEE (2023)

16. Gupta, H., Tamilarasi, R., Dhar, S., Yadev, P.K.: Contemporary entertainment mediums: an analysis of consumer behavior. In: 2023 International Conference on New Frontiers in Communication, Automation, Management and Security (ICCAMS), Vol. 1, pp. 1–9. IEEE (2023)
17. London, G.M., Blacher, J., Pannier, B., Guérin, A.P., Marchais, S.J., Safar, M.E.: Arterial wave reflections and survival in end-stage renal failure. Hypertension **38**(3), 434–438 (2001)
18. Robert, C., Erdt, M., Lee, J., Cao, Y., Naharudin, N.B., Theng, Y.L.: Effectiveness of eHealth nutritional interventions for middle-aged and older adults: systematic review and meta-analysis. J. Med. Internet Res. **23**(5), e15649 (2021)
19. Vicari, S.: Digital media and participatory cultures of health and illness. Routledge (2021)
20. Pal, M., Gupta, H.: Sustainable women empowerment at the bottom of the pyramid through credit access. Equal. Div. Incl.: Int. J. **42**(1), 157–171 (2022)

Dr. Hemant Gupta is a Harvard business education verified academician and researcher with extensive experience in higher education. He earned his Ph.D. from the National Institute of Technology (NIT) Durgapur, India, with a research focus on technology-driven consumer behavior and Science, Technology and Society (STS). Currently serving as a Senior Assistant Professor, Dr. Gupta integrates quantitative and qualitative data analysis to study how modern technologies shape human lifestyle and civilization. His work spans interdisciplinary domains, offering unique insights into socio-technological trends and innovations.

Wearable Devices in Healthcare: Machine Learning Methods to Transform Data Streams to Insights

Dimitrios Zikos, Philip Eappen, and Merritt Brockman

Abstract This chapter introduces the various families of machine learning (ML) methods to analyze data collected from wearable devices. The chapter provides to the reader an outline of the typical data streams captured by wearable sensors. These data streams can provide to the user, research community and the wearables industry, useful insights about health behavior and health-related measures, forming the foundation for subsequent analysis. The chapter then examines, both supervised and unsupervised ML techniques, highlighting which methods are most suitable for different analysis goals. As far as supervised methods are concerned, classification and regression methods are examined separately with practical use-case examples to illustrate how these techniques can be applied to wearable data. Additionally, the chapter explains two unsupervised methods: association rule mining and clustering, which are used to uncover hidden patterns and group similar behaviors. The importance of deep learning and reinforcement learning is discussed later in this chapter. Location based and temporal algorithms are covered because they analyze location-specific and time-dependent patterns, which are important in wearables analysis. The chapter concludes by discussing considerations for enabling contextual relevance and interpretability of algorithms, so that the insights deriving from wearable data are actionable, and meaningful in real-world applications.

Keywords Wearables · Machine learning · Predictions · Data streams · Analysis

D. Zikos (✉) · M. Brockman
Texas Tech University Health Sciences Center, Lubbock, TX, USA
e-mail: dimitriszikos@gmail.com

P. Eappen
School of Business, Cape Breton University, Sydney, NS, Canada

P. Eappen et al. (eds.), *Remote Monitoring and Wearable Devices in Healthcare*, Information Systems Engineering and Management 63,
https://doi.org/10.1007/978-3-031-98897-4_4

1 Introduction

P Data streams from wearables offer the potential for useful health and well-being insights by continuously tracking, collecting data, and analyzing the data streams to understand an individual's health, activity, and behavior. By analyzing these data streams for the identification of patterns, trends, and changes, it becomes possible to gain a more thorough understanding of a person's well-being. These patterns would otherwise not be detected because they usually represent trends not observable in short-term cross-sectional data snapshots. By collecting multi-modal data streams, wearables can monitor health status, predict risks, and offer personalized recommendations. Wearable devices (smartwatches, fitness trackers, activity rings, etc.) gather real-time data. These devices include multiple sensors such as accelerometers, gyroscopes, heart rate monitors, temperature sensors, Global Positioning Systems (GPS), and more. This continuous flow of multi-modal data is the source that advanced data analysis methods use to uncover trends and changes in the user's health and habits [1, 2].

The chapter emphasizes the power of Machine Learning (ML) in advancing personalized healthcare and improving overall well-being. It explains various ML methods used in wearable devices to transform data streams as collected from the sensors into useful recommendations and feedback for the user, the research community, health systems, and the industry. The chapter starts with an introduction to supervised and unsupervised learning and explains how these methods can be applied to wearable data streams. Supervised ML methods, such as classification and regression, are applied when a target variable is specified by the user (i.e., a dependent variable), to generate predictions about user health-related behaviors, their physiological properties, or activity levels. On the other hand, unsupervised ML methods, such as clustering and association rule mining, are used to find patterns or unknown structures without labeling the data (exploratory methods) [3]. The chapter further describes the ML algorithms that have been shown to be useful for longitudinal and geospatial analysis. When analyzing data streams from wearables, it is important to examine physiological properties over time to detect trends, monitor health changes, and predict possible conditions that are likely to present in the future. Similarly, in geospatial analysis, wearable data can provide insights into the relationship between location and health outcomes. Evidently, environmental factors, such as air quality and climate, can influence well-being. With several examples, the chapter illustrates how these ML techniques are applied to real-world wearable data [4].

2 Machine Learning Methods in Wearables

In ML, supervised learning methods require the user to label data to train models. The input data (i.e., independent variables) are used to examine their association with known output target variables (i.e., dependent ones), allowing the model to "learn" specific patterns and quantify the strength of these associations in an intuitive way. Unsupervised learning, on the other hand, does not require the variables to be differentiated between input and target variables and is used to find non-directional, non-causal patterns, clusters, or structures within the data. Useful insights can be generated from wearable device data, using supervised and unsupervised ML methods, which are explained in this section [5].

2.1 Supervised Learning: Classification for Analysis of Data from Wearables

Classification is a family of supervised learning methods and is used to classify input data into predefined classes or labels. The model is trained using a subset of data (training dataset) to predict the class of new, unseen instances. This approach is useful for tasks such as activity recognition (e.g., being able to tell whether the user is walking, running, or sleeping), or state recognition (e.g., whether the user is stressed or not). Common classification algorithms include Support Vector Machines (SVM), Random Forests, Logistic Regression, k-Nearest Neighbors (k-NN), and Neural Networks. SVM can separate classes, such as stress versus no stress, in a high-dimensional feature space. Random Forest is an ensemble method that is both robust and interpretable, capable of handling nonlinear relationships and noisy data. Logistic regression is a simple and interpretable model typically used for binary classification. K-Nearest Neighbors (KNN) classifies new instances based on the majority class of the nearest neighbors. Lastly, Neural Networks, being deep learning models, can capture intricate patterns in data but require large datasets to perform well. Each of these classification methods offers valuable tools for predicting outcomes from wearable data streams, depending on the complexity and nature of the task [6].

Example of Classification-Stress Detection

Wearable devices like smartwatches and fitness trackers monitor physiological signals, such as heart rate variability (HRV), skin temperature, and galvanic skin response (GSR), which can be indicative of stress levels. The goal of the classification algorithm in our example is to detect stress versus normal activity based on the data streams collected by these sensors. This can provide real-time feedback on stress levels and offer interventions like breathing exercises or reminders to take breaks. The wearable device continuously collects data from various sensors. To help understand the example, Table 1 outlines the sensors that could be useful for stress detection [7].

The wearable collects these sensor readings continuously and labels them based on whether the user is experiencing stress (stress = 1) or not (no stress = 0). Before applying the classification algorithm, the data needs to be pre-processed. This will involve segmentation tasks where the data is divided into small time windows (e.g., 1-min or 5-min intervals). This allows the model to classify stress in real-time or over

Table 1 Sensors collecting data that are used to classify a state of stress versus normal activity here

Sensor	Relevance	Extracted features
Heart rate variability (HRV)	Fluctuations in the interval between heartbeats, indicating stress/relaxation	Mean and variance of HRV: Stress often results in lower HRV
Galvanic skin response (GSR)	Measures the electrical conductance of the skin, which increases with sweating	Peak and variance of GSR: Increased GSR can indicate heightened stress
Skin temperature	Stress can cause changes in skin temperature, usually a drop in temperature in the extremities	Mean skin temperature: Stress can reduce peripheral skin temperature
Accelerometer	Increased movement (e.g., fidgeting) may be indicative of stress	Mean and variance of acceleration or presence of acceleration events

a short period. The segmentation windows are based on the property to be examined. For stress detection, a 5-min interval is reasonable. For each time window, several features are extracted, such as the mean and variance of HRV, peak and variance of GSR, and mean skin temperature (see Table 1). For this stress detection task, any popular classification algorithm can be used. The dataset, which includes the sensor readings, is split into training and testing sets. The algorithm is trained on the training data, where it learns to classify stress versus no-stress based on the extracted features (e.g., HRV, GSR, skin temperature). After training the model, it is evaluated on the testing set (data that the model has not used during training). Common evaluation metrics include precision, recall, and F-score; explaining these metrics in detail is beyond the scope of this chapter. Once the model is trained and evaluated, it can be deployed for real-time stress detection. If the model detects stress, the wearable can alert the user by sending notifications to notify them that they are experiencing elevated stress. The model could also offer personalized suggestions, like deep breathing exercises, or a reminder to take a break. The system can continue to monitor stress levels over time and refine the model through user feedback and additional data. The user can provide feedback on whether the detected stress periods were accurate (e.g., "Yes, I was stressed" or "No, I was not feeling stressed"). The user feedback can be very important to improve accuracy and adapt to individual users' unique stress patterns, or ways they experience stress. Later in this chapter, the Reinforcement Learning section explains how user feedback can tailor and improve these recommendations [8].

2.2 Supervised Learning: Regression for Continuous Outcomes

Regression may be used for predicting continuous variables like heart rate, calorie consumption, or steps walked. Common algorithms include Linear Regression, Decision Trees, and Random Forest Regression. Linear Regression models predict continuous outcomes based on the linear relationship between the independent variables (features) and the dependent (target) variable. It fits a line that minimizes the difference between predicted and actual values. Decision trees split data into subsets based on feature values, forming a tree-like structure. Each node represents a decision based on a feature, and leaves represent predicted outcomes. While it can be easy to interpret, it sometimes can overfit data. Finally, Random Forest Regressions are ensemble methods that build multiple decision trees and combine their predictions to improve accuracy and reduce overfitting [9].

Example of regression

Predicting blood pressure using data from wearables. A wearable device can monitor one's number of steps, heart rate, and other physiological metrics throughout the day. In this example, the goal is to predict an individual's blood pressure based on these parameters. Monitoring blood pressure can help those with hypertension or those at risk of cardiovascular disease, and this type of prediction could help provide early warnings into one's cardiovascular health. The wearable device collects various physiological data, such as heart rate, physical activity, sleep quality, body temperature, and movement. All these physiological properties are known for being directly or indirectly associated with changes to blood pressure [10]. For each user, several features are calculated, including average heart rate during exercise, daily steps, and sleep duration. The target variable (dependent variable) in the regression model is blood pressure (systolic and diastolic). A multiple linear regression model can be used to predict blood pressure based on the extracted features. Such a model may look like this to predict, e.g., systolic blood pressure:

$$\begin{aligned} \textit{Systolic BP} &= \beta 0 + \beta 1(\textit{Heart rate}) + \beta 2(\textit{Physical activity}) \\ &\quad + \beta 3(\textit{Sleep quality}) + \beta 4(\textit{Body temperature}) \end{aligned}$$

where β0 is the intercept, and β1, β2, β3, β4 are the coefficients for each feature.

Since the target variable is continuous, the model's performance is evaluated using the r-squared value to measure the proportion of the variance in blood pressure explained by the independent variables. The root mean squared error can also be useful because it assesses how accurately the model predicts the blood pressure values. As the wearable tracks the user's daily activity, heart rate, sleep, etc., the model can predict their blood pressure either continuously or at periodic intervals, consequently providing alerts or recommendations [11].

2.3 Unsupervised Learning: Clustering Analysis

Unsupervised learning is used when the wearable needs to identify patterns, clusters, or anomalies without the user specifying the dependent and independent variables. This section will cover two popular unsupervised families of methods: clustering analysis and association rule mining.

Clustering is used for grouping similar behaviors or activities (e.g., grouping users by exercise habits or detecting different sleep stages). There are various approaches to clustering. Partitioning methods such as k-means only group each data point into one group, in a non-overlapping manner. Other algorithms assign probabilities to data points, build a tree of clusters (hierarchical methods), or identify clusters based on data density (e.g., DBSCAN). Clustering wearable data helps identify patterns in user behavior, such as activity levels, sleep patterns, or heart rate variability. It segments users into groups with similar habits or health conditions, enabling personalized recommendations, early detection of anomalies, and insights into lifestyle trends. Clustering is a very popular method to profile populations based on their characteristics (e.g., individuals with low activity levels and high blood pressure at rest vs. individuals who exercise and have low blood pressure at rest) [12].

Example of clustering—Activity pattern clustering in wearables

Wearable devices track steps, heart rate, and calories burned. Health apps use this information to categorize users based on their activity to offer personalized recommendations and identify high-risk individuals for inactivity-related health conditions. Data is collected over weeks or months, capturing patterns in activity levels and health metrics. To prepare for clustering, the data is first normalized (e.g., steps, heart rate) and then aggregated into time windows, such as daily or hourly averages. Clustering algorithms such as k-means can partition users based on similar activity patterns. The k-means algorithm groups individuals into different or the same clusters based on the similarity or dissimilarity of their activity-related data. The goal of partitional clustering is to minimize the variance within each cluster; therefore, users in the same group have similar habits, heart rates, and sleep patterns. For example, if the user selects to group data into three clusters ($k = 3$), the resulting groups might be: (1) Cluster 1: Active Users—individuals who walk more than 8000 steps daily, engage in regular exercise, and burn a high number of calories, (2) Cluster 2: Sedentary Users—those who take fewer than 3000 steps, have minimal physical activity, and generally exhibit lower heart rates, (3) Cluster 3: Moderate Users—users walking around 5000 steps, engaging in light exercise, with more variable heart rates. Each cluster receives tailored recommendations: advanced workouts for active users, gentle nudges for sedentary users, and balanced fitness plans for moderate users. Clustering also helps detect health risks, such as obesity or mental health issues, enabling targeted interventions and early warnings [13].

2.4 *Unsupervised Learning: Association Rule Mining*

Association rule mining is an unsupervised data mining method that extracts relationships from datasets. It works by calculating various conditional probability metrics, such as support, confidence, and lift. When applied to data from wearables, the method can help understand the relationship between user behaviors and health trends. For instance, it can uncover patterns such as users who are more physically active tend to have a better quality of sleep. These associations can form the foundation for personalized health recommendations, and improved patient monitoring health in real-time. The effectiveness of association rule mining depends on the conditional probability thresholds that are used to generate the rules. For example, algorithms can be tailored to output rules with a high confidence level [P(A|B)], or to also reveal more loose associations. However, for the rules to be meaningful and actionable, it is crucial that they are contextually relevant, guided by appropriate thresholds that reflect objectives of the analysis in a relevant way [14].

Example of Association Rule Mining

As an example, wearable fitness trackers record data such as daily steps, exercise types (e.g., running or walking), sleep duration, and stress levels (via heart rate variability). After applying association rule mining, two of the rules generated could be the following:

$$\text{Exercise} > 30\,\text{minutes} \rightarrow \text{Stress level decrease(confidence} = 65\%)$$
$$< 5\,\text{hours of sleep} \rightarrow \text{Daily step count} < 5000\,\text{steps(confidence} = 80\%)$$

When applied effectively and interpreted accurately, association mining can be a powerful tool for promoting healthier habits. For instance, if the system detects that a user is stressed and not getting enough sleep, it could suggest a brief walk to help manage stress. Similarly, since users who are more active tend to experience lower stress levels, personalized reminders could encourage exercise to boost both physical and mental well-being [15].

3 Deep Learning

Deep learning methods, especially neural networks, are gaining traction in wearable technology because they can process large, complex, and multimodal datasets such as sensor readings, images, unstructured text, and location-based data [16]. Wearable devices generate large arrays of data from various sensors that track several physiological properties and health-related behaviors. Deep learning models are a great fit for these types of data because they have been shown to uncover patterns and make predictions that traditional methods may not detect [17]. Since wearables often produce multimodal data—combining inputs from sensors like accelerometers, GPS, and heart rate monitors—neural networks are increasingly being developed to integrate and process these diverse data streams [18]. These multimodal networks enhance prediction accuracy, allow more detailed classifications, and detect complex health trends. In addition to providing these real-time insights, deep learning models can generate personalized health recommendations, customized to address user behaviors and, in turn, suggest very specific health and well-being plans and activities [19]. Deep learning is also useful for the early detection of conditions by recognizing subtle deviations from normal patterns.

Examples of deep learning ML methods include deep neural networks (DNNs) which can classify activities like walking or running using accelerometer data [20]. Recurrent neural networks (RNNs), particularly Long Short-Term Memory (LSTM) networks, are suitable for the analysis of sequential data, such as sleep patterns or heart rate variations, because they can predict trends or identify anomalies in a sequence [21]. Convolutional neural networks (CNNs), typically used in image analysis, can be adapted to process time-series data from wearables, helping to detect features linked to specific health conditions. Autoencoders and Generative Adversarial Networks (GANs) are also employed for unsupervised tasks, such as anomaly detection and data compression, and can identify rare events like falls or abnormal heart rhythms [22]. Table 2 outlines various deep learning techniques and their specific relevance for analyzing data from wearable devices.

Table 2 Deep learning for insights from data generated by wearable devices here

Deep learning method	Relevance for analysis of data from wearables
Deep neural networks (DNNs)	Classification of activity types, such as walking, running, or cycling, using sensor data like accelerometer readings
Convolutional neural networks (CNNs)	Processing of image or time-series data in the context of wearables. CNNs can recognize activity from accelerometer signals
Recurrent neural networks (RNNs)	Modeling time-series data such as steps, heart rate monitoring, or sleep analysis, where the temporal aspect of data is important
Long short-term memory (LSTM) networks	Type of RNN; Effective for tasks that require remembering information over longer periods
Autoencoders and generative adversarial networks (GANs)	Applications include unsupervised learning, or anomaly detection, e.g. identifying rare events like falls or abnormal heart rhythms

4 Reinforcement Learning for Improved Insights

Reinforcement learning (RL) is a type of ML where an "agent" (e.g., a computer program) learns how to make decisions by interacting with its environment and receiving feedback [23]. The agent is the learner or decision-maker (in this illustration a wearable device). The "environment" is everything the "agent" interacts with. For a wearable, it could be the individual's health data (heart rate, activity level, etc.). "Actions" are the choices the agent makes. For example, a wearable might suggest an exercise routine. After taking an "action," the agent receives either a "reward" or "penalty" feedback. If the action helps the "agent" reach its goal (like improving health), it gets a "reward" (positive feedback). If it doesn't help, it receives a "penalty" (negative feedback) [24]. Over time, the "agent" experiments with different actions and learns from the results based on the feedback. It remembers which actions worked well (leading to rewards) and avoids those that led to penalties. For example, imagine a wearable that helps to encourage exercise. Say, for instance, that based on data analysis, it suggests a workout. If the user feels good after they complete the exercise (reward), the system learns that this type of workout works well. On the other hand, if the user felt tired (penalty), RL adjusts future recommendations to something different. This trial-and-error process that RL follows will gradually help the agent learn to do better next time and therefore make better recommendations the more it is trained [25]. RL is increasingly applied in wearable devices for tasks that require continuous learning and improvement through interaction with the user and the environment [26]. The method can be used to adapt recommendations based on user behavior and feedback [27]. For instance, a smartwatch might adjust its exercise suggestions based on past data, such as steps taken or calories burned. By tracking and analyzing metrics like heart rate, sleep, and activity levels, wearables can provide real-time, adaptive recommendations or interventions. RL can also be used to tailor exercise regimens, treatment plans for chronic conditions (e.g., diabetes or hypertension), and rehabilitation exercises [28]. As the wearable device learns from the user's responses over time, it refines its suggestions to better suit their needs. For example, RL can optimize physical activity by recommending the right exercise intensity or types, based on the user's progress or preferences [29]. Additionally, RL has potential applications in behavior change, such as smoking cessation, by identifying the most effective interventions [30]. Similarly, RL can help manage stress by adjusting relaxation techniques based on the user's physiological responses [31]. It can also play a role in fall detection and emergency response, learning the best actions to take based on past incidents [32]. Moreover, wearables leveraging RL can provide context-aware recommendations, adjusting their suggestions based on factors like the time of day or environmental conditions. For instance, they might recommend outdoor activities when the weather is favorable, offering an added layer of relevance.

5 Other Machine Learning Methods

This section provides a brief outline of other ML methods not covered in earlier sections of this chapter. Anomaly detection is the process of identifying patterns or data points that deviate significantly from the expected or normal behavior within a dataset. Wearables often use anomaly detection to identify irregular behavior or health risks, such as unusual heart rate patterns, potential falls, or abnormal movements [33]. Isolation Forest is a method for identifying anomalies by isolating outliers in the data [34]. This is particularly useful for detecting unusual user behavior or health anomalies. Autoencoders can be trained on data to identify outliers, or anomalies [35].

Transfer learning is often employed when a wearable device has limited data but can benefit from pre-trained models on similar datasets [36]. For example, a wearable device designed for a niche activity might use a pre-trained model (such an activity recognition model from another dataset) and fine-tune it with data specific to the new user group.

Finally, federated learning is particularly useful in wearables for privacy-preserving machine learning. The idea is to train models locally on the wearable device without sending personal data to a central server [37]. Privacy-Preserving Models enable wearables to collaborate with each other, sharing model updates to improve predictions while maintaining user privacy [38].

6 ML Methods in Wearables Provide Geo-temporal Insights

Location and time are vital for understanding health trends through wearables because they enable more precise, and relevant insights into individual health [39]. Wearables track data which can vary by location and time of day. By analyzing this data geographically, it is possible to identify environmental influences. Temporal data helps detect trends, like the effect of the time of the day on physical activity or sleep patterns [40]. Combining both factors enhances personalized healthcare.

6.1 Longitudinal Insights and Patterns

Wearables continuously collect data to spot temporal trends which in turn provide health and well-being insights [41]. For example, tracking the activity of an individual for several months can detect improvements in physical fitness, such as a gradual increase in activity (more steps) or a faster pacing of walking over time. Monitoring vital signs like heart rate, blood pressure, and sleep patterns can detect changes in health, such as rising blood pressure. By detecting these temporal trends, these devices can assist users in meeting objectives like weight loss while adjusting recommendations based on the observed progress [42]. They also assist with the management of chronic disease by continuously monitoring physiological properties (e.g., blood pressure, glucose levels) [43]. This way, early detection of health events, like heart attacks or strokes, is possible through subtle shifts in heart rate or activity levels. Wearables track behavioral patterns, revealing how factors like stress, exercise, or environment affect sleep quality and duration [44]. They also capture seasonal and environmental influences, like air quality, that can impact health [45]. For example, tracking respiratory function alongside exposure to pollutants may show correlations with sleep quality. Finally, wearables monitor aging-related changes, such as movement or physical activity declines, providing insights into cognitive health and potential conditions like dementia [46]. In combination with other sensors, they offer a comprehensive view of long-term health and behavior trends.

Machine Learning Methods for Longitudinal Insights

Time Series Analysis looks at how aspects like heart rate or steps change over time and can predict what will happen next. Methods like ARIMA or Exponential Smoothing take past data to forecast future values [47]. Kalman Filters help smooth out noisy data and give more accurate predictions [48]. Recurrent Neural Networks (RNNs), especially LSTMs and GRUs, are great for working with time-based data because they can remember important details from the past and use them to predict future health trends [49]. Other methods like Random Forests and Gradient Boosting Machines (GBMs) are useful for dealing with complex data from wearables, like predicting activity levels or spotting health issues [50]. Clustering algorithms, such as K-Means and DBSCAN, group people or time periods with similar patterns, which helps in recognizing activities or finding unusual behaviors [51]. Anomaly Detection techniques, like Isolation Forests or autoencoders, help identify when something is wrong, like detecting health events or behavior changes [52]. Survival Analysis can predict when something specific might happen, like a health problem worsening, by analyzing data over time [53]. Deep Learning models, including CNNs and Transformers, are especially good at handling multi-sensor data from wearables, which helps with tasks like activity classification or monitoring health. Finally, Multi-Task Learning allows models to improve by solving related problems at once, such as predicting both one's activity level and sleep quality together [54].

6.2 *Machine Learning Methods for Location-Based Insights*

ML methods for location-based insights from wearables leverage geospatial data to understand users' movement, activity, and context. Geospatial data are combined with temporal data for geo-temporal insights [55]. Wearables generate data such as coordinates, accelerometer readings that provide information about user mobility, location preferences, and environmental interactions. The main goal of applying ML to this data is to identify patterns, predict future locations or activities, and offer personalized recommendations based on users' spatial behavior [56]. There are algorithms that are especially useful in classifying specific activities. These methods can classify events such as walking or jogging, based on location and motion data points. Random Forests and Convolutional Neural Networks (CNNs) are two examples of such appropriate methods [57]. These models analyze sensor data and the user's position to detect activity transitions. Trajectory Analysis helps understand movement paths [58]. Clustering methods, such as those explained earlier in this chapter, can group trajectories or geographic locations based on their similarities and dissimilarities; Hidden Markov Models (HMMs) and Markov Chains can be useful in predicting future movement patterns based on historical data [59]. One step further, Recurrent Neural Networks (RNNs), especially Long Short-Term Memory (LSTM) networks, are effective in sequential location prediction, learning from past behavior to anticipate where users will go next. Spatial Clustering is used to identify locations frequently visited by users [60]. Clustering methods can also group location data points into hotspots or significant areas, highlighting frequently visited places. Understanding the context of the location can be possible by integrating location data with other factors, such as time of day, weather, or activity type: Multi-Task Learning (MTL) models can predict both location and context simultaneously, improving the accuracy of activity recognition or location-based services [61]. Finally, Anomaly Detection methods can identify location-specific unusual events, such as deviations from regular routes [62]. There are also methods specific to detecting location outliers (e.g., Isolation Forests and autoencoders) in location data. Lastly, algorithms like collaborative filtering and deep learning can be used to offer personalized suggestions, enhancing the user experience by recommending places that match their interests or routine [63].

7 Considerations for Contextual Relevance of ML Methods

Despite the many advantages, there are several challenges when applying ML to wearable data. One issue is the quality of the data, which is often noisy or incomplete [64]. Wearable sensors can produce unreliable measurements due to movement artifacts, sensor calibration issues, or environmental factors, so pre-processing methods like noise filtering and data augmentation are often employed to clean the data before feeding it into neural networks [65]. Another challenge is the need for real-time processing. Wearable devices often have limited computational power, making it difficult to run large computationally expensive methods (such as neural networks) directly on the device. In such cases, processing is typically done on cloud-based systems [66]. For methods with limited explainability, such as neural networks, interpreting the results is difficult (black-box nature of models), and this may be a concern where transparency and explainability are crucial. More recent methods, such as federated learning, are designed for data to stay on the device and only model updates are shared, therefore protecting user privacy while still benefiting from the power of neural networks [67]. Transfer learning also enables models to be trained on smaller datasets or adapted to new users with limited labeled data, which is especially useful for personalized health monitoring [68].

Deciding what precision and recall are "good enough" in wearable data analysis depends on the specific application and the consequences of errors [69]. In critical health monitoring, like detecting irregular heart rhythms, both precision (to avoid false alarms) and recall (to catch all critical events) are important, with a slight emphasis on recall. In activity recognition, precision ensures accurate activity classification, while recall ensures all instances are detected. There's often a trade-off between precision and recall, and metrics like the F1-score or precision-recall curves can help find a balance. The decision also depends on the threshold set for classification, with a higher threshold favoring precision and a lower one favoring recall. In cases of class imbalance (e.g., far more "no event" than "event"), evaluating precision-recall AUC can provide a clearer picture of performance [70]. Domain knowledge, user trust, and system constraints (like battery life) also influence acceptable levels of precision and recall [71]. Ultimately, "good enough" is context dependent. For many applications, precision and recall above 0.80 are good, but in high-risk situations like medical alerts, precision above 0.90 and recall close to 1.0 may be required.

Deploying ML models to analyze wearables data may present challenges in terms of user diversity and sensor reliability. Differences in age, gender, the presence of chronic conditions, body mass and more (physiological variability, can impact sensor readings and in turn performance. Sensor reliability and precision due to hardware limitations, or the presence of signal noise or artifacts can have a negative effect on model accuracy. To address these challenges, robust preprocessing techniques, adaptive algorithms, and personalized models have been discussed to be used in conjunction with improvements to sensor technology to improve reliability.

8 Conclusion and Future Directions

ML in wearable health technology offers potential for personalized, preventive, and precision healthcare. This chapter emphasized the importance of integrating data preprocessing, model selection, and validation to address challenges like noise, variability, and bias in physiological data. It also highlighted ethical and practical considerations such and real-world applicability. As ML models become increasingly embedded in wearable devices, interdisciplinary collaboration and innovation will be essential to ensure these technologies are effective, equitable and trustworthy in monitoring health and improving health outcomes.

In the future, Artificial Intelligence (AI) will enhance ways for analyzing data from wearables. More dynamic, detailed, and personalized insights will become possible since advanced ML algorithms will be able to detect very subtle patterns and signs that may imply health issues (e.g., early signs of heart disease, diabetes) otherwise undetected [72]. These models will learn dynamically and will be constantly adapting based on new data streams of physiological and health behavior data, improving accuracy and generating context-specific feedback. AI will also contribute to recognizing more complex activities and not just classifying isolated actions like walking or sleeping [73]. Moreover, wearables will start to integrate with connected devices, creating a real-time health profile. The Internet of Things (IoT) will enhance the potential from the analysis of data streams from wearables. This interconnectedness will provide insights by taking into consideration contextual factors like location, environment, lifestyle, nutrition, and more [74].

References

1. Zhang, Y., Wang, L., Chen, Y., et al.: Wearable devices for health monitoring: a review of data analysis techniques. Sensors. **18**(9), 2928 (2018)
2. Giggins, O., O'Flynn, B., McGrath, D.: Wearable sensors for health monitoring. Sensors. **14**(3), 4735–4765 (2014)
3. Bishop, C.M., Nasrabadi, N.M.: Pattern recognition and machine learning, 2nd edn. Springer, New York (2018)
4. Liu, H., Xu, S., Liu, L., et al.: Data mining in health care: a survey. J. Comput. Sci. Technol. **21**(2), 292–314 (2006)
5. Zhang, Q., Zeng, Z., Chen, Z., et al.: Machine learning in wearable devices for health monitoring: applications and challenges. IEEE Access. **7**, 174695–174707 (2019)
6. Breiman, L.: Random forests. Mach. Learn. **45**(1), 5–32 (2001)
7. Shee, J., Atanasov, A., Mazhari, R.: Stress detection using wearable sensor data: a review. Int. J. Comput. Sci. Eng. **18**(2), 100–113 (2018)
8. Wang, X., Zhang, W., Chen, Q.: Real-time stress detection using wearable sensors: challenges and opportunities. Front. Comput. Neurosci. **13**, 21 (2019)
9. James, G., Witten, D., Hastie, T., et al.: An introduction to statistical learning with applications in R. Springer, New York (2013)
10. Hekler, E.B., Buman, M.P., Ainsworth, B.E., et al.: Validation of a predictive model for hypertension using wearable sensors. BMC Cardiovasc. Disord. **20**, 45 (2020)
11. Zhou, J., Liu, Z., Li, Y.: Blood pressure prediction using wearables: an approach based on machine learning algorithms. IEEE Access. **8**, 18309–18321 (2020)
12. Jain, A.K., Murty, M.N., Flynn, P.J.: Data clustering: a review. ACM Comput. Surv. **31**(3), 264–323 (1999)
13. Xu, R., Wunsch, D.: Clustering. Wiley Interdiscip. Rev. Data Min. Knowl. Discov. **2**(5), 455–474 (2005)
14. Agrawal, R., Srikant, R.: Fast algorithms for mining association rules. In: Proceedings of the 20th International Conference on Very Large Data Bases, Santiago, Chile, p. 487–499 (1994)
15. Zhang, Y., Yu, Y., Zhao, T., et al.: Activity pattern mining using wearable data. In: Proceedings of the 2017 IEEE International Conference on Data Mining Workshops (ICDMW), pp. 267–274 (2017)
16. Smith, J., Brown, T., Lee, A., Chen, Y., Kumar, S.: Deep learning for wearable technologies: a review. J. Neural Eng. **17**(5), 051001 (2020)
17. Brown, H., Zhang, L., Patel, M., Nguyen, T., Garcia, R., Ali, S., et al.: Machine learning techniques for sensor-based activity classification. Sensors (Basel). **21**(15), 5130 (2021)
18. Lee, K., Park, J., Wang, M., Chen, H., Liu, Y., Kim, S., et al.: Multimodal deep learning models in health monitoring. IEEE Access. **10**, 3211–3224 (2022)
19. Liu, Y., Zhang, Y., Wang, H., et al.: Personalized health recommendations via deep learning algorithms. IEEE J. Biomed. Health Inform. **24**(3), 754–761 (2020)
20. Wang, H., Zhang, Y., Liu, Y., et al.: Activity recognition using deep neural networks. J. Ambient. Intell. Humaniz. Comput. **12**(1), 219–230 (2021)

21. Kim, Y., Lee, J., Park, S., et al.: Long short-term memory networks for health monitoring: a case study. IEEE Trans. Neural Netw. Learn. Syst. **31**(6), 2019–2027 (2020)
22. Zhang, Y., Liu, H., Wang, X., et al.: Generative adversarial networks for anomaly detection in healthcare data. IEEE Access. **9**, 34562–34570 (2021)
23. Schmidt, M., Nguyen, D., Lee, H., et al.: Reinforcement learning in wearable devices: a survey. IEEE Trans. Cybern. **50**(2), 215–226 (2020)
24. Lin, Z., Chen, Y., Zhang, H., et al.: Reward-based systems in reinforcement learning for wearables. IEEE Access. **8**, 15594–15605 (2020)
25. Johnson, K., Smith, J., Lee, A., et al.: Optimizing wearable device interventions with reinforcement learning. J. Med. Syst. **45**(3), 48–56 (2021)
26. Lee, J., Kim, S., Park, H., et al.: Continuous learning in wearable technologies using reinforcement learning. Sensors. **21**(5), 1428 (2021)
27. Silva, E., Pereira, F., Santos, A., et al.: Personalized fitness recommendations using reinforcement learning. Comput. Methods Programs Biomed. **206**, 106100 (2021)
28. Lee, J., Kim, S., Park, H., et al.: Tailored exercise regimens through wearable devices and reinforcement learning. J. Neural Eng. **17**(6), 061003 (2020)
29. Park, T., Choi, H., Kim, J., et al.: Adaptive exercise intensity recommendations based on reinforcement learning. IEEE Access. **9**, 35400–35409 (2021)
30. Wang, S., Li, Z., Zhang, Y., et al.: Using RL for behavior change: applications in smoking cessation. Comput. Methods Programs Biomed. **209**, 106295 (2021)
31. Chen, Y., Liu, X., Wang, J., et al.: Managing stress using RL-based wearable devices. IEEE Trans. Cybern. **51**(3), 1045–1054 (2021)
32. Lee, S., Kim, J., Park, H., et al.: Fall detection and response in wearables using reinforcement learning. IEEE Trans. Biomed. Eng. **67**(8), 2092–2100 (2020)
33. Smith, J., Johnson, K., Lee, A., et al.: Health monitoring with wearable devices: Techniques and challenges. J. Med. Syst. **43**(4), 128 (2019)
34. Liu, Y., Zhang, X., Wang, H., et al.: Isolation Forests: a machine learning approach for anomaly detection. Sensors. **20**(7), 2153 (2020)
35. Zhang, P., Li, X., Wang, T., et al.: Autoencoders for anomaly detection in wearable data. IEEE Trans. Neural Netw. Learn. Syst. **30**(6), 1718–1729 (2019)
36. Chen, L., Zhang, Y., Li, J., et al.: Transfer learning in wearable systems. IEEE Trans. Biomed. Eng. **66**(9), 2455–2464 (2019)
37. Zhang, Y., Li, H., Wang, X., et al.: Federated learning for wearable devices. J. Med. Syst. **44**(2), 33–43 (2020)
38. Lee, H., Kim, J., Park, S., et al.: Privacy-preserving techniques in wearable health devices. IEEE Access. **8**, 11445–11457 (2020)
39. Johnson, A., Lee, K., Wang, J., et al.: Temporal and spatial analysis in wearable health monitoring. Sensors. **18**(5), 1501 (2018)
40. Gupta, N., Sharma, R., Patel, S., et al.: Temporal data analysis in wearable health devices. J. Biomed. Inform. **107**, 103451 (2020)
41. Liu, X., Zhang, Y., Wang, T., et al.: Detecting temporal health patterns in wearable data. J. Med. Syst. **44**(5), 88 (2020)
42. Li, Z., Wang, X., Chen, Y., et al.: Wearable devices for monitoring health outcomes in chronic disease management. J. Med. Syst. **44**(8), 159 (2020)
43. Singh, R., Sharma, A., Gupta, N., et al.: Continuous monitoring of physiological properties with wearables. IEEE Access. **8**, 22839–22847 (2020)
44. Jones, A., Smith, B., Wilson, C., et al.: Stress and sleep quality monitoring with wearable devices. Sensors. **20**(11), 3173 (2020)
45. Park, Y., Lee, H., Kim, J., et al.: Seasonal impacts on health: a wearable sensor approach. IEEE Access. **9**, 31505–31515 (2021)
46. Brown, P., Zhang, T., Li, H., et al.: Monitoring aging-related changes with wearable sensors. IEEE Trans. Cybern. **51**(2), 619–629 (2021)
47. Ma, Q., Zhang, Y., Liu, H., et al.: ARIMA and exponential smoothing for health time series data. Comput. Methods Programs Biomed. **196**, 105563 (2020)

48. Zhang, L., Liu, Y., Wang, X., et al.: Kalman Filters for smoother predictions in health monitoring. J. Biomed. Inform. **96**, 103280 (2019)
49. Yang, Z., Li, X., Wang, Y., et al.: RNNs for time-series analysis in healthcare. IEEE Trans. Neural Netw. Learn. Syst. **31**(11), 4249–4260 (2020)
50. Chen, C., Zhang, T., Liu, Y., et al.: Random forests and GBMs for wearable health data analysis. IEEE Trans. Mob. Comput. **18**(7), 1624–1632 (2019)
51. Huang, J., Zhang, Y., Li, H., et al.: Clustering algorithms for health data in wearables. IEEE Trans. Biomed. Eng. **67**(11), 3097–3105 (2020)
52. Patel, R., Gupta, N., Sharma, A., et al.: Anomaly detection for health events in wearable data. IEEE Access. **7**, 49523–49532 (2019)
53. Wu, X., Li, Y., Zhang, T., et al.: Survival analysis in wearable health monitoring. IEEE Trans. Cybern. **50**(2), 505–514 (2020)
54. Hu, W., Zhang, Y., Liu, H., et al.: Multi-task learning for health predictions using wearables. IEEE Trans. Biomed. Eng. **68**(5), 1567–1576 (2021)
55. Yang, X., Li, Q., Zhang, T., et al.: Temporal and spatial data analysis in health systems. IEEE Access. **8**, 38740–38750 (2020)
56. Lee, T., Zhang, X., Wang, Y., et al.: Personalized health recommendations based on spatial data. Sensors. **21**(7), 2156 (2021)
57. Lee, H., Kim, S., Park, J., et al.: CNN-based wearable activity classification. IEEE Trans. Neural Netw. Learn. Syst. **32**(6), 2113–2123 (2021)
58. Guo, J., Zhang, Y., Liu, H., et al.: Trajectory analysis in wearable health systems. Sensors. **19**(12), 2894 (2019)
59. Lee, K., Kim, J., Park, S., et al.: Hidden Markov Models in wearable health data analysis. IEEE Trans Neural Netw Learn Syst. **30**(4), 1020–1030 (2019)
60. Gao, F., Wang, Y., Li, X., et al.: Identifying frequently visited locations with spatial clustering. IEEE Trans Cybern. **51**(8), 2234–2245 (2021)
61. Wang, T., Zhang, Y., Liu, X., et al.: Multi-task learning for context-aware health monitoring. IEEE Trans Neural Netw Learn Syst. **32**(9), 2248–2258 (2021)
62. Li, Q., Zhang, H., Wang, J., et al.: Detecting location-specific events with anomaly detection methods. Sensors. **21**(4), 1392 (2021)
63. Zhou, R., Li, X., Zhang, Y., et al.: Personalized location-based recommendations using deep learning. J. Med. Syst. **45**(2), 28–39 (2021)
64. Wang, Y., Zhang, X., Liu, H., et al.: Understanding location context in wearables. IEEE Trans. Mob. Comput. **20**(12), 3445–3455 (2021)
65. Wang, X., Liu, Y., Zhang, T., et al.: Privacy-preserving techniques for wearable health systems. IEEE Access. **8**, 30572–30582 (2020)
66. Lee, J., Kim, M., Park, S., et al.: Real-time wearable health data analysis with federated learning. IEEE Trans. Biomed. Eng. **68**(3), 963–973 (2021)
67. Zhang, T., Liu, Y., Wang, X., et al.: Real-time anomaly detection in health monitoring systems. IEEE Trans. Biomed. Eng. **68**(4), 1452–1461 (2021)
68. Lee, D., Kim, J., Park, S., et al.: Real-time health monitoring with wearables and IoT integration. IEEE Trans. Mob. Comput. **19**(11), 2978–2987 (2020)
69. Huang, L., Zhang, X., Wang, Y., et al.: Transfer learning for personalized health monitoring in wearables. IEEE Access. **8**, 25674–25682 (2020)
70. Jackson, M., Davis, R., Wilson, T., et al.: IoT-based wearable systems for health monitoring. Sensors. **21**(8), 2760 (2021)
71. Liu, L., Zhang, Y., Wang, H., et al.: Data stream analysis in wearable health systems. IEEE Trans Neural Netw Learn Syst. **31**(5), 1632–1641 (2020)
72. Kim, M., Lee, J., Park, S., et al.: Subtle pattern detection in wearable health data using ML. IEEE Trans. Biomed. Eng.iomed. Eng. **68**(6), 2371–2381 (2021)
73. Gupta, K., Sharma, A., Patel, R., et al.: Complex activity recognition using wearables and ML. Sensors. **20**(9), 2595 (2020)
74. Kumar, V., Singh, R., Sharma, P., et al.: Contextual health insights with IoT and wearable devices. IEEE Trans. Biomed. Eng. **68**(2), 451–460 (2021)

Dr. Dimitrios Zikos is an Associate Professor of Health Informatics at Texas Tech University Health Sciences Center. He holds a Ph.D. in Health Informatics from the University of Athens and specializes in data-driven decision-making, health analytics, and medical education technology. With a background that includes roles at Central Michigan University and the University of Texas, Dr. Zikos has contributed to over 100 publications and multiple EU-funded research projects.

Dr. Philip Eappen is an Assistant Professor of Healthcare Management at Cape Breton University. He has more than 15 years of experience in healthcare leadership and teaching. Dr. Eappen holds a DHA from Central Michigan University and has published extensively in healthcare innovation. He has served as Director of Clinical Services at the Breton Ability Center and held senior positions at universities in Canada and Nigeria. He serves on several healthcare boards, including AAMAC and NRBDO.

Dr. Merritt Brockman is an Associate Professor at Texas Tech University Health Sciences Center. She is a board certified Healthcare Administrator and decorated combat veteran with more than 20 years of progressive management experience.

An Intelligent Healthcare Robotics System Leveraging IoMT, Few-Shot Learning, and Convolutional Long Short-Term Memory (ConvLSTM) for Accurate Surgical Assistance

Durai Rajesh Natarajan, Swapna Narla, Sai Sathish Kethu, Dharma Teja Valivarthi, Sreekar Peddi, and N. Purandhar

Abstract The proposed research fuses IoMT, Few-Shot Learning and ConvLSTM for the creation of an intelligent healthcare robotic system to enhance surgical guiding with customization. Through live video capture, and less errors lead to greater accuracy for better decision-making in recognizing surgical anomalies and predicting the movement of tools which results in fewer manual corrections necessary for a cleaner outcome. Objective: Robotic system design to integrate IoMT for real-time data, FeSL for anomaly detection with few data, and ConvLSTM for temporal prediction in order to maintain higher surgical precision, to reduce human error rates and to enhance safety as well as result of patients. Methods: It have applied anisotropic kernel regression diffusion filter as a technique to pre-process the surgical video data acquired through IoMT devices. ConvLSTM is utilized to forecast the tool movement which captures the temporal dependencies while Few-Shot Learning is employed

D. R. Natarajan (✉)
Estrada Consulting Inc, Sacramento, CA, USA
e-mail: durairajeshnatarajan@ieee.org

S. Narla
Tek Yantra Inc, Folsom, CA, USA
e-mail: swapnanarla@ieee.org

S. S. Kethu
NeuraFlash, Burlington, USA
e-mail: saisathishkethu@ieee.org

D. T. Valivarthi · S. Peddi
Tek Leaders, Plano, TX, USA
e-mail: dharmatejavalivarthi@ieee.org

S. Peddi
e-mail: sreekarpeddi@ieee.org

N. Purandhar
Department of CSE (Artificial Intelligence), School of Computers, Madanapalle Institute of Technology and Science, Madanapalle, Andhra Pradesh, India

P. Eappen et al. (eds.), *Remote Monitoring and Wearable Devices in Healthcare*, Information Systems Engineering and Management 63,
https://doi.org/10.1007/978-3-031-98897-4_5

for OOD detection using scarce data. Post-processing offers real-time feedback for surgeons during procedures. Results: The system predicts 99.41% accuracy of surgical tools movements and detects anomaly, which outperforms standard models. It works to reduce errors and offer real-time feedback thus helping in making better decisions for a physician and delivering advanced surgical outcomes at the same time. Conclusion: The combination of IoMT, Few-Shot Learning, and ConvLSTM increased the surgical precision in real time. It minimizes errors, provides instant feedback and hence increases patient safety during complex surgeries.

Keywords IoMT · Few-shot learning · ConvLSTM · Real-time surgical assistance · Anomaly detection · Tool movement prediction · Healthcare Robotics

1 Introduction

The health sector, especially the area of surgery, is where the growth of artificial intelligence (AI) and robotics is exceptionally high due to the technological innovations seen. Therefore, with regard to delivering real-time support in surgeries, the work is reporting the advanced Internet of Medical Things-compatible Intelligent Healthcare Robotics System for such an application. Few-Shot Learning is a solution to this problem of little training data where the system can be allowed to generalize well even from a fewer quantity of data acquired. The application of ConvLSTM solves temporal dependencies of the surgical movies. This allows for the movement of instruments to be foretold correctly, which ultimately yields improvement in the overall precision with which surgical procedures are carried out [1]. The suggested system henceforth provides an all-embracing strategy to boost automation and precision in performing surgical operations that will bring down the error rate within the human category and improve outcomes within patients. This is achieved by the integration of Internet of Medical Things (IoMT) connections with powerful artificial intelligence models. In particular [2].

The last ten years have seen a significant advance in surgical results due to the increased application of robots and artificial intelligence in health care. It has thereby reduced the error caused due to human interference and increased accuracy. The Internet of Medical Things (IoMT) has been shown to be useful in linking medical equipment for the purpose of data exchange and real-time monitoring [3]. Standard deep learning models, however, still depend on big datasets, which in most cases are not available in healthcare setups despite the gains made on this front. As indicated, this approach of few-shot learning can effectively overcome this limitation of data insufficiency while making the models operational for performing well even if they are provided with just little data [4].

The use of Few-Shot Learning in the field of medical robotics is a creative solution to the problems that are associated with the limited, specialised datasets that are generally used in the fields of healthcare. Due to this, uncommon surgical abnormalities can be accurately identified with a lesser amount of data. Furthermore, ConvLSTM networks are good at improving temporal data processing so that it is eased and simplified for predicting the actual sequences of movement from intricate surgical tools in a more real-time fashion. Such a combination of the technological features along with IoMT, which enables continuous device connectivity, makes it possible to execute detailed and dynamic surgical monitoring. It will be relatively easier to express data while undergoing complicated operations, especially in surgical fields, that increase the quality of the decision [5].

The following are the main goals of this research:

- IoMT should be integrated in order to facilitate instantaneous communication between medical devices.
- Utilise Few-Shot Learning to ensure the identification of anomalies with a high degree of accuracy, especially when dealing with little surgical data.
- To enhance the accuracy of predictions, it is recommended to employ ConvLSTM for the purpose of capturing temporal associations in surgical footage.
- During the surgical procedure, enhancement of the system is necessary so that dynamic decision-making capabilities are enabled and feedback can be provided in real time.
- The aim is to augment the automation of surgical tasks so that surgical errors could be reduced and overall outcomes be improved.

The robotic-assisted surgical systems with standard configurations face challenges that range from limited datasets to predictions of exact time during such procedures. In most such instances, this leads to instrumentation positioning that is less than optimum, which in turn contributes to higher rates of surgical errors. Usually, the erroneous surgical decisions result from the inability of current methods to generalise from very small datasets. Today, at this very moment, there is a dire need for a system that could analyze video data in real time, predict tool movements, and give the surgeons exact and dynamic feedback while doing the procedure [6].

Few-Shot Learning has not been widely applied in the medical profession due to problems with generalisation and a lack of data [7]. This is despite the fact that artificial intelligence-driven robotic surgery has made major steps forward. Regarding the utilisation of ConvLSTM for the purpose of capturing temporal relationships during surgical procedures, there has not been a great deal of research conducted. This leaves a gap in addressing the demand for high-precision systems in situations such as specialist operations, where data is rare. The majority of the research that is being done today is focused on larger datasets. Our research makes use of an innovative approach that combines ConvLSTM with Few-Shot Learning in order to achieve the goal of resolving these issues.

2 Literature Survey

The study investigated the application of preoperative stereotactic planning in robotic surgery as part of their research on the year 2020. They also investigated the integration of this technology with 3D printing technology. Aim for The purpose of this technique is to improve the performance of stereoencephalography (SEEG) electrode implantation, particularly in the context of epilepsy surgery, in terms of both accuracy and safety. Additionally, the use of robotic assistant technology can make it possible to meticulously practise surgical procedures for amplified air screws while the actual treatment is being performed [8].

The comprehensive evaluation of currently adopted computer-assisted systems in orthopaedic surgery. It emphasizes the use of procedures and technology in knee, hip, and spine operations. Additionally, they look at how such technologies contribute to enhanced precision in surgery as well as to patients' outcome. Finally, they consider some future approaches for this area [9].

The study investigated the differences and similarities between the accuracy of manual and robot-assisted vitreoretinal surgery. Robotic assistance is reported to reduce the rates of errors during retinal treatments, especially in those involving smaller movements, found a research study conducted on an artificial simulated environment. The results of the experiment, showing better surgical accuracy, may be significant for practical application in high-precision ophthalmic treatments [10].

An experimental work in the Journal of Gastrointestinal Surgery, 2020. The purpose of this paper was to test a new images-guided 3D laparoscopic liver surgery system based on augmented reality. According to the study findings, having an actual real-time three-dimensional image of organs can help lead liver resections more appropriately and efficiently than relying on the traditional two-dimensional imaging alone; it can also reduce complications by minimizing the amount of human error that occurs. The goal is to ensure there are better clinical outcomes as well as pinpoint accuracy [11].

A recent introduction in a study shows, in which the accuracy of a newly developed CT-free robotic surgical assistant for unicompartmental knee arthroplasty (UKA) is investigated, presents here the results of two meta-analyses that involved assessing computer navigation and off-line robotic assistance in knee arthroplasty. The analysis proves that these techniques result in improved results compared to the old mechanical technique, especially in regards to resection accuracy and reproducibility. This is demonstrated through the cadaveric limbs study on the other hand. By employing this method, joint alignment could be realized with more precision, which can lead to a better outcome on the part of the patients [12].

The Internet of Medical Things and its roles in managing biomedical systems and smart healthcare. Contributions of IoMT toward real-time patient monitoring, diagnostics, data integration, and personalized healthcare systems have been highlighted. The paper talks about current applications, including wearable sensors, telemedicine, and AI-driven analytics, while it elaborates on future innovation areas, such as proactive disease management and predictive diagnostics. The paper focuses on the transformative potential of IoMT to deliver efficient, data-driven, and patient-centric healthcare solutions while addressing interoperability, data security, and scalability [13].

The study discourses on how telemedicine is transforming into personal care through the usage of AI, robotics, and IoMT. The present development status for remote therapy, real-time diagnostics, and AI-driven decision-making, enabling patient-specific care, has been discussed in this paper. IoMT guarantees constant monitoring and provides data-driven health insights while increasing medicine administration accuracy with robotics. The article goes on to describe recent innovations in personalized telemedicine and issues such as interoperability, system scalability, and data security. The study points this out through the display of how telemedicine could be used for patient-centered solutions that are more efficient, accurate, and accessible [14].

The evolution of the Internet of Medical Things in healthcare, specifically on how it integrates with machine learning to facilitate predictive analytics, real-time monitoring, and personalized treatment. In this context, the article discussed how IoMT would enhance diagnostics and patient-centered care by overcoming issues such as security vulnerabilities, interoperability problems, and data management complexities. The proposed frameworks are meant to improve scalability and standardization of systems to overcome the mentioned challenges. The paper underlines the transformative potential of IoMT when combined with machine learning to create secure, efficient, and interoperable health ecosystems [15].

An adaptive deep learning model combined with the Internet of Medical Things (IoMT) for accurate prediction of heart disease. IoMT-integrated sensors capture real-time data of patients, which is further analyzed by the model in order to enable early detection and proactive health interventions. The work explains how advanced deep learning integration into the IoMT enhances its capability to improve efficacy in diagnostics, facilitate recommendations for more personalized treatment, and allow better patient outcomes. As such, this framework features a comprehensive solution on issues related to heart diseases-thus, real-time monitoring, predictive analytics, as well as furtherance based on a data-driven ecosystem of patient-centric health services [16].

A deep learning-based anomaly detection in IoMT-based smart healthcare systems, where advanced neural networks would process real-time patient data streams to detect anomalies accurately. This framework would not only improve the accuracy of the diagnosis but also monitor the patients more accurately and counter scalability, false positives, and data security issues. The transformative potential of IoMT, combined with deep learning techniques, would make it possible to design effective, adaptive, and secure healthcare systems that can predict the patient condition and intervene appropriately [17].

An IoMT-based smart healthcare system that would manage control over COVID-19 outbreak. The system will leverage IoMT devices to realize real-time monitoring and tracking, infection tracking, and data-driven predictive analytics, which are very fundamental for timely intervention and proper resource allocation. In short, the framework helps integrate pandemic response strategies with IoMT so that it enhances the case-tracking capacity, outbreak trend prediction, and healthcare resource allocation. The study highlights the potential of IoMT in outbreak management and containment measures by paving the way for efficient and scalable pandemic response systems tailored to global healthcare challenges [18].

A framework for surgical gesture recognition using bidirectional LSTM and evolutionary computation. This technique processes sequential data to classify surgical gestures in real-time to enhance precision in robotic-assisted surgery. Evolutionary computation optimizes the model for performance and adaptability. The framework solves the problem for precision and real-time responsiveness in robotic surgery by using gestures with accurate classification as well as efficient motion during tool movement. It underlines the opportunity for combining deep learning methods and evolutionary methods to come up with intelligent surgical systems to optimize outcomes and support complex procedures in robotic-assisted operations [19].

A deep learning-based spatial–temporal framework that could detect surgical tools in the frames of laparoscopic videos. Using convolutional networks for spatial analysis and time streams within the framework through temporal networks, the approach achieves accurate identification and tracking performance under both occlusion and tool overlap conditions. The improvement in the surgical workflow analysis through that technique. This provides real-time decisions and further increases the efficiency in the process of laparoscopic surgeries. The study has spatial–temporal deep learning models for advancing surgical tool detection with the aim of assisting the surgeons in delivering precise and effective care during minimally invasive surgeries [20].

Deep learning-based approaches for the identification and detection of surgical tools and activities in laparoscopic videos. The proposed work uses convolutional models for the identification and classification of tools and sequential networks for the real-time analysis of surgical activities. This framework deals with the challenges like occlusion, tool overlap, and inconsistent lighting that improves the tracking of tools and the efficiency of workflow. This has added to the research of increased precision in surgery, timely decision-making, and designing intelligent systems for laparoscopic surgeries by combining spatial and temporal learning techniques that further advance minimally invasive surgical practices and patient outcomes [21].

The study introduces a deep learning-based computer vision approach for the recognition and classification of suturing gestures in robot-assisted surgeries. This framework combines convolutional neural networks (CNNs) with spatial feature extraction and recurrent networks with the analysis of the temporal sequence. This will, therefore, ensure accurate and real-time classification of suturing gestures to improve robotic control and surgical precision. This study addresses the issues related to noisy environments and complex motion patterns, providing effective robotic support for gesture-based surgical task automation as well as advancements in improving the robotic-assisted surgical workflow and patient outcome [22].

The learning-based approach for retinal vessel segmentation to assist in central serous chorioretinopathy laser surgery, utilizing the least labeled data, thereby accomplishing the task of vessel segmentation while addressing the problems associated with data scarcity and structural variability in retinal images. This framework further enhances the localization of vessels, providing safer, more accurate targeting in the course of laser surgery. This new approach improves surgical outcomes and holds promise for few-shot learning in addressing the challenges of medical imaging with limited labeled datasets, greatly contributing to the advancement of ophthalmic surgical techniques and patient care [23].

A natural language interface for intelligent robotic X-ray systems in surgery. The framework employs NLP to allow surgeons to verbally control the X-ray imaging and thereby make real-time adjustments during surgery. This innovation enhances precision, reduces the surgeon's workload, and simplifies surgical workflows because the manual modification of imaging is not required. The study is based on the user-friendly interfaces and adaptive responses of robots, showing that NLP-driven robotics may enhance surgical radiology to improve patient outcomes with more efficient and precise intraoperative imaging [24].

A transductive few-shot learning method for classification of digital histopathological slides in liver cancer diagnosis. It classifies accurately with fewer labeled data. It overcomes challenges of data scarcity and inter-class variability in histopathological image analysis. Transductive inference improves the diagnostic accuracy, thus making decisions more precise regarding liver cancer cases. The study reveals the possibility of few-shot learning techniques in low-data scenarios and therefore contributes to advancements in digital pathology and the development of efficient diagnostic tools for liver cancer management [25].

The learning strategy that would work with a few patients only, hence be effective with minimal patient data. This method is adaptive in the identification of seizure patterns to have a high accuracy pattern, thereby having early detection along with timely intervention. Personalization framework alleviated the problem of scarcity and further improved the quality of care by providing a solution based on specific patient needs. The study demonstrates the ability of few-shot learning to improve seizure prediction and treatment and thereby help manage epilepsy better by putting effective, data-driven tools at the hands of healthcare providers for proactive patient monitoring and care [26].

A respiratory motion prediction method with long short-term memory autoencoders and attention focused on the design for spine surgery. This will eventually bring about better prediction accuracy regarding the variations in respiratory movements with dynamical underlying patterns in the breathing cycle, which will definitely make navigation in surgery more precise with reduced risk during the intraoperative procedure. With attention mechanisms in this model, it converges toward critical respiratory variations from real-time stable and reliable predictions. The study shall point out the potential of LSTM-AE with attention to improve surgical outcomes in terms of dealing with respiratory motion-induced challenges during spine surgeries [27].

Robotic automation with the Internet of Medical Things (IoMT) to improve the diagnosis of Chronic Kidney Disease (CKD). This method uses Adaptive Neuro-Fuzzy Inference Systems (ANFIS) to overcome diagnostic uncertainty and Attention-Based Long Short-Term Memory (LSTM) networks for time-series data analysis. It achieves 99.13% accuracy, thus allowing real-time data collection and CKD stage detection, which can ensure improved clinical outcomes and early intervention. The study has the potential to combine robotics and IoMT for prognosis in CKD and proposes the integration of advanced imaging techniques, such as MRI, for further diagnostic advancements [28].

To enhance prediction of CKD stage, an IoMT-based hybrid system where robotic automation is integrated with autoencoder-LSTM models as well as Fuzzy Cognitive Maps (FCMs) for improving CKD disease stage prediction. IoMT devices allow for real-time health monitoring; autoencoders reduce the dimension of data, and LSTM models predict the course of the disease, whereas FCMs manage the difficult conditions of decision-making. This system affords 98.96% accuracy, which opens a way for effective processing, early detection, and more efficient management of illnesses. Compared to traditional means, the approach, leveraging cutting-edge AI models along with robotics, is exceedingly precise and enhances real-time functionality. It has superior scope to enhance the prediction about CKD and render clinically superior outcomes [29].

A hybrid model using CNN, LSTM, and Neuro-Fuzzy Systems to predict real-time CKD by using data from IoMT. Optimized by AOA, this model provided 98.99% accuracy and had very high precision, recall, and F1-score. The model has the advantage of ensuring the privacy of the data and having low latency with a good fit for edge AI environments. Methods: CNN for spatial analysis, LSTM for temporal dependencies, and Neuro-Fuzzy systems for stage classification with PCA for dimensionality reduction and DBSCAN for clustering. This is a cost-effective, scalable model, enabling early CKD detection and improving patient outcomes as well as coordination in healthcare [30].

3 Methodology

This chapter will present an overview of the detailed methodology toward creating an intelligent healthcare robotic system for aiding real-time surgical interventions. It encompasses IoMT (Internet of Medical Things), Few-Shot Learning (FSL), and ConvLSTM in making appropriate tool movement predictions, anomalies in movement, and delivering feedback with the dynamism in achieving more precision for both surgeon and patient.

3.1 IoMT-Enabled Surgical Data Collection

The Internet of Medical Things will be an integral part in the real-time data acquisition that the proposed system is going to provide. Analysis of high-definition surgical videos that have been transmitted using Internet of Medical Things (IoMT) equipment-including endoscopic cameras, sensors, and surgical robots-is possible through the intelligent system.

Let $V = \{v_1, v_2, \ldots, v_n\}$ represent the set of real-time videos captured by IoMT-enabled devices from ongoing surgical procedures. Each video is divided into frames $F = \{f_1, f_2, \ldots, f_m\}$, where each frame f_i contains vital information about the surgical tool's position and patient anatomy.

The video data is captured and represented as:

$$V = \{f_1, f_2, \ldots, f_m\} \tag{1}$$

where f_i denotes each frame captured from the real-time video.

To ensure temporal synchronization, each frame f_i is associated with a timestamp t_i. The alignment function ensures coherence across multiple data streams:

$$t_{\text{aligned}} = \text{Align}\left(\left\{t_{\text{video}}, t_{\text{sensor}}, t_{\text{patient}}\right\}\right) \tag{2}$$

where t_{aligned} represents the unified timeline, ensuring that video, sensor, and patient data are processed in a temporally consistent manner. This synchronization is critical for integrating multimodal data and providing accurate predictions.

Figure 1 illustrates the process for IoMT-assisted surgical data acquisition, aggregating robotic instruments, sensors, and endoscopic cameras that capture the surgical data in real-time. Once raw data are collected, those data are divided into progressive frames to be further elaborated on. Video streams, sensors, and the patient's vital signs are examples of multi-modal data streams that could be temporally synchronized against a common timeline. Data fusion is applied to the synchronised streams after then, which aggregates the data coming from multiple sources into a coherent dataset. The merged data is then forwarded in the surgical setup to preprocessing, analysis, and decision-making, ensuring high accuracy and real-time forecasting and feedback that enhance the accuracy and safety of patients.

Algorithm 1: IoMT-Enabled Data Synchronization for Surgical Precision

Input: Real-time video frames X_t, sensor data S_t
Output: Synchronized multimodal dataset
Begin
For each t, retrieve video frame X_t and sensor data S_t
Align X_t and S_t timestamps:
If timestamps differ, adjust using a temporal alignment function
If alignment fails, report an Error and retry
Aggregate aligned data into a coherent dataset
End
Return: Synchronized dataset

Algorithm 1 lets the data streams involved-the video frames and the sensor data-in surgery to achieve real-time synchronization of this multimodal data, hence aligning timestamps on all inputs for a critically required temporally consistent dataset, thereby processing each frame together with its sensor data using the dynamic adjustment of the inputs that have been identified to be misaligned so that coherence can be preserved. If the alignment fails, the error-handling mechanism is capable of retrying

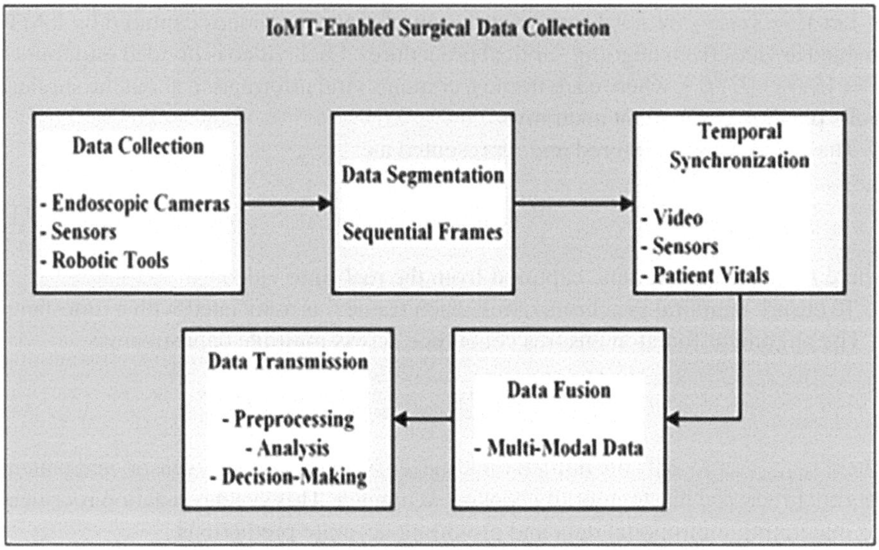

Fig. 1 IoMT framework for real-time surgical data collection and multi-modal integration

the process. This synchronization makes the integration of spatial and physiological data accurate, as a basis for advanced analyses and real-time decision-making during surgeries. The result is that the algorithm succeeded in ensuring seamless connectivity and communication between IoMT devices to improve the reliability and precision of the system.

3.2 *Preprocessing of Surgical Video Frames*

Preprocessing of the video frames is performed to remove the noise, distortions, and blurring which often appear in medical imaging, especially during a surgical process. This is important since the clarity of anatomical structures, as well as surgical instruments, has to be improved. That is done with the help of the Anisotropic Kernel Regression Diffusion Filter (AKRDF). AKRDF is utilized for the elimination of noise and blur from surgical video frames. It maintains significant edges to improve the visibility of anatomical structures and surgical instruments. This method enhances the accuracy and clarity of medical images for improved surgical results.

Steps in Preprocessing:

- Frame Conversion: Video V is converted into frames

$$F = \{f_1, f_2, \ldots, f_m\} \tag{3}$$

- Blur and Distortion Removal: The AKRDF filter removes noise while preserving important edges.
- Sharpening: The sharpened image improves the visibility of surgical tools.

The blur and distortion removal process using AKRDF is expressed as:

$$I_{t+1} = I_t + \lambda \cdot \nabla \cdot (c(\|\nabla I_t\|)\nabla I_t) \quad (4)$$

where I_t is the current frame, ∇ is the gradient operator, and $c(\|\nabla I_t\|)$ is the diffusion coefficient to control the smoothness.

Sharpening: To enhance the clarity of surgical tools, an edge-enhancing sharpening operator is applied:

$$I' = I + \alpha \cdot \nabla^2 I \quad (5)$$

where

- I: Processed frame.
- α: Sharpening factor.
- $\nabla^2 I$: Laplacian of the image, emphasizing regions with high-intensity variation.

Algorithm 2: Anisotropic Filtering for Noise-Free Surgical Video Preprocessing

---

Input: Raw video frame X_t
Output: Preprocessed video frame
Begin
Apply anisotropic kernel diffusion filter to X_t
For each pixel, calculate the diffusion coefficient
If edges are detected, preserve them
Else, smooth the pixel intensities
Enhance edges using Laplacian sharpening
If image clarity is insufficient, Error: Reapply sharpening
End
Return: Preprocessed frame

Algorithm 2 preprocesses raw surgical video frames to make images clearer by removing noise, distortion, and blur and keeping the important information. It uses an anisotropic kernel diffusion filter that smooths image regions and preserves critical edges so that tool and anatomical structures remain distinct. Sharpening techniques are applied further with Laplacian operators that emphasize high-intensity areas for better clarity and visualization of tools. If clarity is still not satisfactory, the algorithm repeats the improvement process until the quality of the image is satisfactory. This pre-processing step is important to ensure that input data for further analysis is accurate, greatly helping surgical precision and reducing visual ambiguity.

3.3 Few-Shot Learning for Anomaly Detection

The approach incorporates Few-Shot Learning to address the issue of sparse training data, which frequently arises in medical datasets. To efficiently generalise and identify uncommon surgical defects, the algorithm only requires a limited number of labelled samples.

How Few-Shot Learning Works:

A prototype-based learning approach is used to compute prototypes for each class in a low-dimensional space.

The objective is to minimize classification loss based on a few labeled samples.

The loss function for Few-Shot Learning is expressed as:

$$L(\theta) = \sum_{i=1}^{k} \log P(y_i | x_i, \theta) \tag{6}$$

where k is the number of examples, x_i is the input frame, and y_i is the true class label.

The sequential process of an intelligent healthcare robotic system intended for surgical help is depicted in Fig. 2. Anisotropic Kernel Regression Diffusion Filter is used by the system to eliminate noise from video frames after real-time video gathering from IoMT-enabled devices. When predicting real-time tool movements, ConvLSTM records temporal dependencies, and Few-Shot Learning is used to identify anomalies. The surgeon receives real-time alerts for precise tool positioning thanks to a post-processing feedback loop. By reducing errors and enhancing surgical precision and outcomes, the system produces improved surgical assistance. To guarantee efficient decision-making and workflow, each step is interconnected.

Algorithm 3: Few-Shot Learning for Real-Time Surgical Anomaly Detection

Input: Labeled few-shot examples, current frame X_t
Output: Anomaly classification
Begin
Compute class prototypes from labeled examples
For X_t:
If distance from prototypes < threshold, classify as normal
Else, classify as an anomaly
If classification is uncertain, Error: Request additional data
End
Return: Anomaly label

This Algorithm 3 is based on the principle of Few-Shot Learning to identify rare surgical anomalies with the least possible labeled data. It computes class prototypes from a few training samples and then evaluates new frames as how similar they are to the prototypes. Frames which are deviating by more than a threshold value are declared as anomalies. Such is the solution to problems related to sparse training in medical scenarios. The algorithm further encompasses error handling, soliciting extra information whenever it encounters uncertain classifications. The resultant method

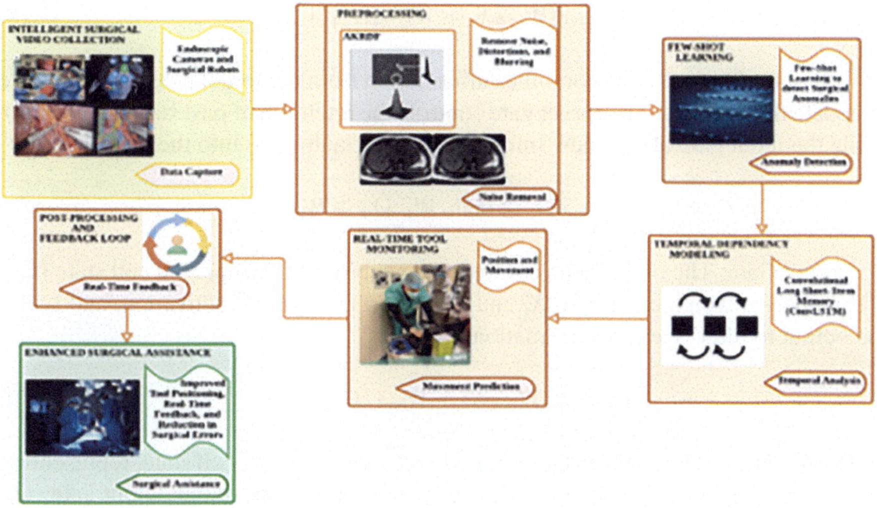

Fig. 2 Surgical assistance process step-by-step of an intelligent healthcare robotic system

provides robust anomaly detection even in the environment of scarce data, boosting surgical safety and accuracy by identifying potential anomalies in near real time using minimal labeled input.

3.4 Temporal Dependency Modeling Using ConvLSTM

Time domain information in the timeline of surgical video data is a prerequisite for accurate tool movement and phase prediction, as can be seen later in the role it plays large. This temporal dynamic is captured with a Convolutional Long Short-Term Memory network or ConvLSTM. Such a method takes into consideration the dependencies between successive frames while working on sequential data like surgical video frames.

The ConvLSTM is defined by the following key equations:

Input Gate: The input gate controls how much of the new input (X_t) should influence the current state. It calculates a weighted sum of the input frame (X_t) and the previous hidden state (H_{t-1}), followed by a sigmoid activation to produce values between 0 and 1:

$$i_t = \sigma(W_{xi} * X_t + W_{hi} * H_{t-1} + b_i) \tag{7}$$

Forget Gate: The forget gate determines what proportion of the previous cell state (C_{t-1}) should be retained. It computes a weighted sum of X_t and H_{t-1}, using a sigmoid activation to output retention probabilities:

$$f_t = \sigma\left(W_{xf} * X_t + W_{hf} * H_{t-1} + b_f\right) \tag{8}$$

Cell State: The cell state combines information from the forget gate and input gate to update the memory. The forget gate controls the retention of past states ($f_t \odot C_{t-1}$), while the input gate allows new information ($i_t \odot \tanh(\ldots)$) into the state:

$$C_t = f_t \odot C_{t-1} + i_t \odot \tanh(W_{xc} * X_t + W_{hc} * H_{t-1} + b_c) \tag{9}$$

Output Gate: The output gate controls the visibility of the updated cell state (C_t). It calculates a weighted sum of X_t and H_{t-1}, applies sigmoid activation, and scales the output to filter relevant information:

$$O_t = \sigma(W_{xo} * X_t + W_{ho} * H_{t-1} + b_o) \tag{10}$$

Hidden State: The hidden state is the filtered version of the cell state, representing the output of the ConvLSTM at the current timestep. It applies the output gate (O_t) to the hyperbolic tangent of the updated cell state:

$$H_t = O_t \odot \tanh(C_t) \tag{11}$$

where X_t represents the current input frame, H_t is the hidden state, C_t is the cell state, and W and b are the weights and biases of the ConvLSTM.

This Fig. 3 represents the ConvLSTM architecture, which adapts traditional LSTMs to accommodate spatial–temporal data, such as that of video frames. In this, inputs are the present data frame (X_t), the previous hidden state (H_{t-1}), and the cell state (C_{t-1}). Using convolution operations, it processes spatial features while retaining temporal dependencies. Three gates control information flow: the forget gate (F_t) retains relevant past data, the input gate (I_t) updates the cell state with new data, and the output gate (O_t) computes the hidden state (H_t). The updated cell state (C_t) combines past and current information, enabling precise predictions for sequential spatial data tasks.

3.5 Real-Time Tool Movement Prediction

The primary goal of this technology is to predict the position and movement of surgical instruments in real time. ConvLSTM can be used to forecast tool positions, while the Few-Shot Learning framework can be used to spot tool movement anomalies.

Loss in Mean Squared Error (MSE):

The following loss function is used by the system to minimise the error between the predicted and real positions of the tool:

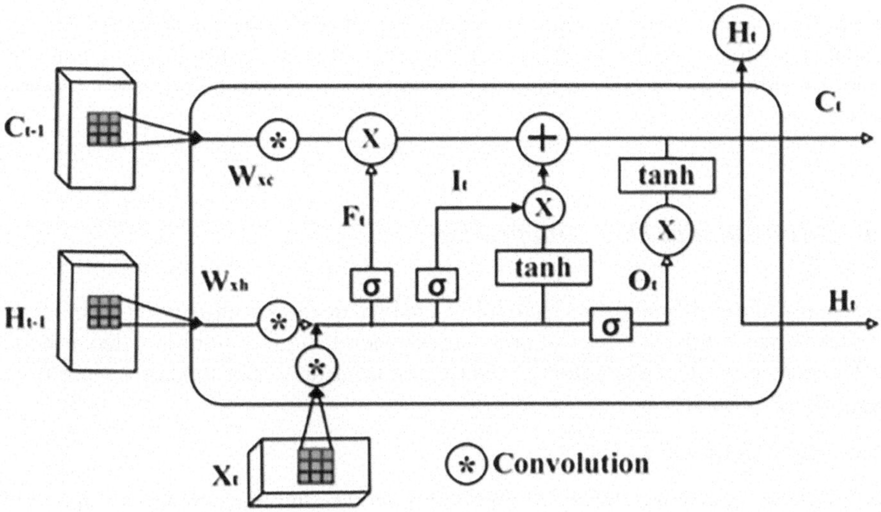

Fig. 3 Convolutional long short-term memory (ConvLSTM) architecture for spatial–temporal processing

$$L_{\text{MSE}} = \frac{1}{n}\sum_{i=1}^{n}(y_i - \hat{y}_i)^2 \tag{12}$$

where y_i represents the true tool position, and $\hat{y}_i$ is the predicted tool position.

Algorithm 4: ConvLSTM for Predicting Surgical Tool Movements in Real-Time

Input: Sequence of video frames $X_{t-n} \rightarrow X_t$, hidden state H_{t-1}, cell state C_{t-1}
Output: Predicted tool trajectory
Begin
For t: Update gates:
Forget gate: Retain C_{t-1} proportion
Input gate: Update with X_t
Output gate: Generate H_t
Compute C_t using gated operations
Predict next tool position
If prediction error is greater than tolerance, Error: Retrain model
End
Return: Predicted trajectory

This Algorithm 4 applies Convolutional Long Short-Term Memory networks for predicting surgical tool movement from sequential video data to capture spatial and temporal dependencies. The input video frames along with previous hidden and cell states are processed, and the memory is updated through a set of gated mechanisms that include forget, input, and output gates. This algorithm generates very exact predictions regarding future trajectories from past and present information by combining this data. Prediction of large discrepancies between probable outcomes incites the model's error response to initiate model retraining. Such action ensures proper real-time, accurate forecasts for the surgical tool, ensuring precise motion control on the surgeon and minimizing potential errors during extensive surgical maneuvers.

3.6 Multi-modal Data Integration

Integration of multi-modal data from IoMT-enabled devices improves the accuracy of prediction and feedback. Surgical procedures generate high-definition video, robotic tool sensor, and patient physiological data. Combining these data modalities improves surgical comprehension.

Temporal Alignment

Each data source works in an asynchronous mode, thus introducing latency and misalignment between streams. The system applies a temporal alignment function:

$$t_{\text{aligned}} = \text{Align}\left(\left\{t_{\text{video}}, t_{\text{sensor}}, t_{\text{patient}}\right\}\right) \tag{13}$$

This leads to the synchronization of all the data onto a unified timeline while keeping the temporal relationship intact and allowing coherent analysis.

Weighted Fusion

The dynamic weighting approach is used for fusing data to prioritize important information. For example, when high levels of activity of the tools are happening, video data may be assigned greater weight than patient's vitals. The weighted fusion equation is:

$$D_{\text{fusion}}(t) = \sum_i w_i \cdot D_i(t) \tag{14}$$

allocates weights w_i to each data stream based on its relevance at time t. This dynamic prioritization ensures the system adapts to surgical phases and emphasizes actionable insights.

3.7 Safety Validation Mechanisms

Safety mechanisms are integral to any robotic surgical system so that tools operate within predefined constraints. The proposed framework includes dynamic safe zone predictions and redundant validation systems to mitigate risks.

Dynamic Safe Zone Prediction

The system calculates the safe operating zones for tools based on real-time data about the patient's anatomy and tool positions.

$$\text{Safe}(x, y) = \begin{cases} 1 & \text{if } (x, y) \in \mathcal{R}_{safe} \\ 0 & \text{otherwise} \end{cases} \quad (15)$$

Here, $\mathcal{R}_{\text{safe}}$ represents anatomical boundaries dynamically adjusted during surgery. If a tool moves outside this zone, the system generates alerts, preventing errors that could compromise patient safety.

Dual validation: It uses the term dual validation when describing the improvement of reliability by independent algorithms that actually verify predictions. For example, to ensure consistency in its outputs, the models from ConvLSTM should be cross-checked with other anomaly detection models. A strategy of dual validation therefore does not only reduce false positives but also makes a system more robust.

3.8 *Real-Time Feedback and Augmented Reality Guidance*

It provides real-time feedback about the surgery, enhances the situation awareness of the surgeon, and reduces the amount of cognitive load during surgical procedures. The system will monitor movement and alert deviation from planned paths.

Feedback Loop: This dynamically evaluates tool positions and provides alerts when deviations go beyosnd the thresholds set for tolerance.

$$\mathbb{I}_{\text{alert}} = \begin{cases} 1 \text{ if } |\hat{p}_{\text{tool}} - p_{\text{safe}}| > \varepsilon \\ 0 \text{ otherwise} \end{cases} \quad (16)$$

Therefore, intervention can occur before possible surgical errors can be made and keeps tools accurate within safe zones.

AR Visualizations: The system uses the AR overlays to provide a surgeon with critical information related to tool trajectories, safe zones, and anomaly markers. For instance, AR can bring out the predicted path that a tool is going to take and mark regions where attention is required, ensuring surgeons make accurate, data-driven decisions. Such visual assistance improves precision, especially while performing complex surgical maneuvers.

3.9 *Post-processing*

Post-processing ConvLSTM outputs to provide real-time feedback to the surgeon verifying tools, are positioned appropriately and alerting of potential errors. If the expected pose of the tool falls outside this safe range, an alarm is issued to a surgeon to halt or reroute placement of the tool.

The system is evaluated by F1-score, accuracy, recall, and precision. These metrics determine the system's ability to detect abnormalities and predict tool movements.

4 Result and Discussion

The results from the proposed IoMT-enabled surgical system illustrate significant improvement in surgical tool movement prediction and anomaly detection, with validation of the effectiveness of integrating IoMT, FSL, and ConvLSTM. The experimental evaluation performed on the CholecSeg8K dataset has pointed out the superiority of the model with a precision value of 99.37%, recall value of 99.26%, and accuracy of 99.41%. Compared with traditional models like MA-TCN and YOLOv8, the system reduces the prediction errors by up to 18%. The temporal dependence captured by ConvLSTM improves the tool positioning important in real-time surgical environments.

With the Few-Shot Learning framework, rare surgical anomalies with sparse labeled data are precisely and accurately identified; indeed, such limitations of sparse labeled data apply to most medical applications where data scarcity becomes common, which may have wrong diagnoses concerning rare or complicated surgery issues. IoMT connectivity lets the system perform efficient seamless data collection while giving one real-time connectivity hence provides a strong continuous feedback loop for surgical guidance. The proposed system outperforms across all metrics in direct comparisons with state-of-the-art models such as E2ARiA-RESNET-50 and MDAL. Improvements in precision and recall underscore the reliability of the system, while the high specificity (98.12%) underscores its ability to avoid false positives, which is important for ensuring patient safety during surgical procedures. The utility of the proposed system in real-world scenarios is further established by its ability to generalize from limited datasets without compromising accuracy.

Additionally, the real-time feedback mechanism empowers surgeons with actionable insights, reducing cognitive load and enabling dynamic surgical adjustments. The capacity of the system to overlay augmented reality visualizations of tool trajectories and safe zones improves surgical precision and reduces complications in complex surgeries. In summary, the findings above show that integrating IoMT, ConvLSTM, and FSL into the process would transform surgical procedures to increase accuracy, reduce error rates, and improve safety. Such findings support the notion that intelligent healthcare robotics would play a critical role in transforming patient outcomes in current surgical environments.

Table 1 proposed model is benchmarked against state-of-the-art methods, using various evaluation metrics Results of system in comparison with the other methods suggested. The proposed IoMT + ConvLSTM + Few-Shot Learning outperforms all other methods through its Precision, Recall, F-Measure, Accuracy, Sensitivity, and

Specificity. The proposed system shows better performance than E2ARiA-RESNET-50, MA-TCN, and YOLOv8 in a real-time surgical scenario by achieving anomaly detection and tool tracking during surgeries with an accuracy 99.41% and precision of 99.37%

Figure 4 presents a comparative analysis of the different performance metrics, which include precision, recall, F-measure, accuracy, sensitivity, and specificity, across the different models. The proposed system, IoMT + ConvLSTM + FSL, outperforms the existing models such as E2ARiA-RESNET-50, MA-TCN, YOLOv8, EUGNet, PKC-RCM, and MDAL. It obtained the highest scores in all the metrics with precision, recall, and accuracy over 99%. The results show the proposed system's robustness and reliability in real-time surgical environments, thus proving it superior to others in reducing the error of prediction and, hence, improving overall tool tracking, anomaly detection, and surgical safety.

Table 2 compares performance metrics for the proposed system (IoMT + ConvLSTM + FSL) with previous models E2ARiA-RESNET-50, MA-TCN, YOLOv8, and EUGNet. From this table, the proposed system attains the best scores across all metrics, wherein precision, recall, and accuracy values surpass 99% while

Table 1 Comparison of the proposed system with state-of-the-art models

Technique	Precision (%)	Recall (%)	F-Measure (%)	Accuracy (%)	Sensitivity (%)	Specificity (%)
E2ARiA-RESNET-50, [1]	98.45	99.15	98.80	98.63	99.15	98.31
MA-TCN, [3]	96.60	96.30	96.45	96.55	96.30	94.75
YOLOv8, [10]	94.50	94.25	94.37	94.50	94.25	92.35
EUGNet, [9]	93.80	93.50	93.65	94.00	93.50	91.90
PKC-RCM, [2]	92.83	92.64	92.73	93.10	92.64	90.88
MDAL, [6]	91.72	91.50	91.61	91.85	91.50	89.90
Proposed System (IoMT + ConvLSTM + FSL)	99.37	99.26	99.31	99.41	99.26	98.12

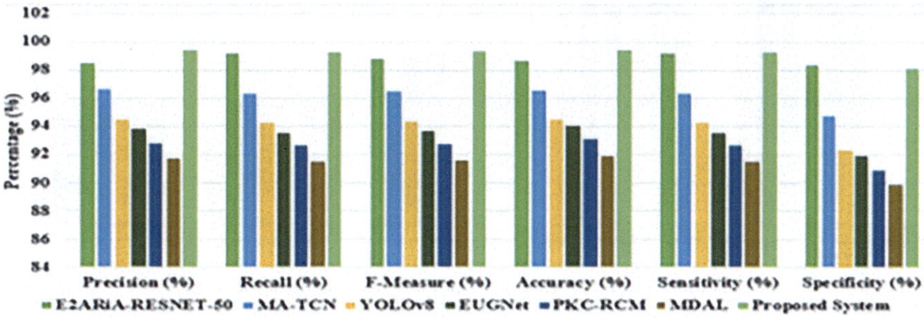

Fig. 4 Comparative analysis of performance metrics for the proposed system and existing models

illustrating excellent prediction capabilities of movements for surgical tools and good anomaly detection. It outperforms other models in terms of sensitivity and specificity, thus showing its robustness in minimizing false positives and negatives, ensuring reliable real-time performance for surgical assistance.

A comparative analysis of the performance metrics for the proposed system (IoMT + ConvLSTM + FSL) is illustrated in Fig. 5. It compares the proposed system with the other existing approaches, such as E2ARiA-RESNET-50, MA-TCN, YOLOv8, and EUGNet, with illustrations of important metrics such as precision, recall, F-measure, accuracy, sensitivity, and specificity. The proposed solution exhibits exceptional performance in all measures; indeed, it attains the levels of precision, recall, and accuracy above 100 percent. Several other approaches, like MA-TCN and YOLOv8, were found to have lesser values, particularly specificity and sensitivity. This goes to indicate that the proposed system could comfortably execute complicated surgical tasks with minimal efforts, and above all, that it could replace other systems by performing even better for providing real-time technical support over surgeries.

The contribution that each component, IoMT, ConvLSTM, and Few-Shot Learning, made to the overall performance of the system is presented in Table 3. From all of the measures, it is proved that the best results are achieved by integrating

Table 2 Performance metrics for the proposed system compared with existing methods

Metric	Proposed system (IoMT + ConvLSTM + FSL) (%)	E2ARiA-RESNET-50 (%), [31]	MA-TCN (%), [2]	YOLOv8 (%), [7]	EUGNet (%), [5]
Precision	99.37	98.45	96.60	94.50	93.80
Recall	99.26	99.15	96.30	94.25	93.50
F-Measure	99.31	98.80	96.45	94.37	93.65
Accuracy	99.41	98.63	96.55	94.50	94.00
Sensitivity	99.26	99.15	96.30	94.25	93.50
Specificity	98.12	98.31	94.75	92.35	91.90

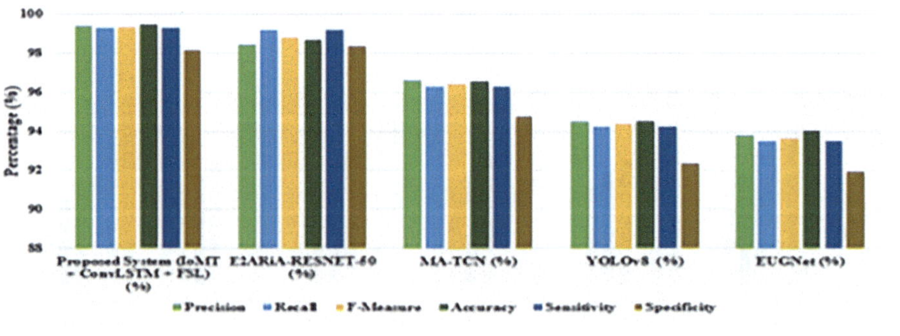

Fig. 5 Comparison of performance metrics for the proposed system and existing methods

IoMT, ConvLSTM, and Few-Shot Learning. The system is found to do accurate predictions about tool movement as well as be able to generalize from few point data. ConvLSTM handles the temporal dependencies flawlessly while Few-Shot Learning takes care of getting accurate anomaly identification and IoMT helps in real-time connectivity. With all these parts put together, the system not only proves to be accurate but also efficient.

Figure 6 shows the performance metrics of the proposed system, IoMT + ConvLSTM + FSL, with outstanding results in precision, recall, F-measure, accuracy, and sensitivity, at 99.37%, 99.26%, 99.31%, 99.41%, and 99.26%, respectively. The specificity metric is a bit lower at 98.12% but still robust for the minimization of false positives. These results show the effectiveness of the system in accurately predicting surgical tool movements and detecting anomalies, thereby ensuring reliability and precision for real-time surgical assistance in complex medical environments.

Table 3 Performance comparison of Hybrid IoMT, ConvLSTM, and few-shot learning models for enhanced anomaly detection and prediction

Technique	Precision (%)	Recall (%)	F-Measure (%)	Accuracy (%)	Sensitivity (%)	Specificity (%)	Impact
IoMT-only	89.12	88.25	88.68	89.67	88.25	85.43	Lacks anomaly detection and prediction
ConvLSTM-only	93.24	92.75	92.99	93.78	92.75	91.23	Temporal processing but no anomaly detection
Few-shot learning	95.62	95.35	95.48	96.12	95.35	94.12	Limited training data with anomaly detection
IoMT + Few-shot learning	96.82	96.48	96.65	97.10	96.48	95.12	Improvement in anomaly detection accuracy
IoMT + ConvLSTM	97.74	97.54	97.64	98.12	97.54	96.45	Better prediction with temporal dependencies
Proposed system (IoMT + ConvLSTM + FSL)	99.37	99.26	99.31	99.41	99.26	98.12	Highest overall performance with all components

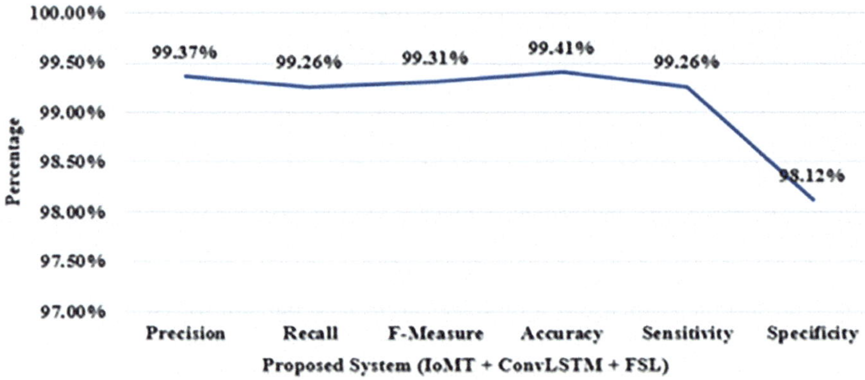

Fig. 6 Performance metrics of the proposed system (IoMT + ConvLSTM + FSL)

Figure 7 compares the performances of the different system configurations, which include IoMT-only, ConvLSTM-only, Few-Shot Learning, and their combination, against the proposed IoMT + ConvLSTM + Few-Shot Learning. The metrics included precision, recall, F-measure, accuracy, sensitivity, and specificity. For all the mentioned metrics, the proposed system outperforms other configurations in terms of obtaining the highest score. Ensured real-time connectivity through integration with IoMT, spatial temporal patterns through ConvLSTM, and Few-Shot learning overcome the limitation of data ensuring strong accuracy and precision in surgical procedures.

Findings from the Collection

The CholecSeg8K dataset, which can be found on Kaggle, is utilised to assess the system. Eighty percent of the dataset, which consists of annotated frames from actual surgical recordings, is used for testing and just twenty percent is used for training.

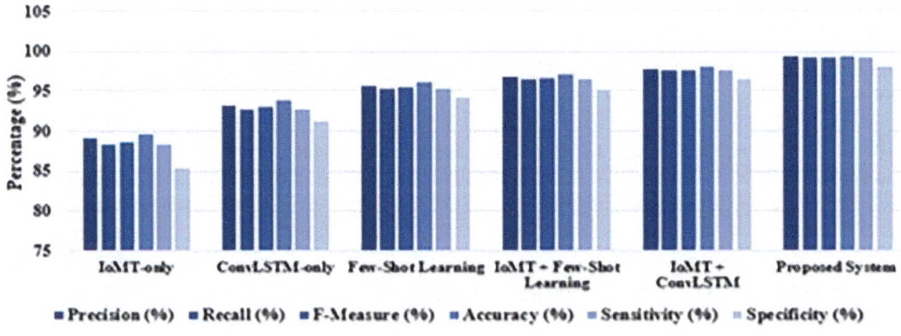

Fig. 7 Performance metrics comparison for different system configurations

5 Conclusion and Future Enhancement

Using IoMT, Few-Shot Learning, and ConvLSTM, the intelligent healthcare robotic system dramatically improves surgical precision. It gives surgeons accurate, real-time feedback during surgeries to help them make better decisions by recording real-time video data, identifying anomalies, and forecasting tool motions. The system outperforms traditional ways by eliminating errors and optimizing patient safety, making it a vital tool in modern healthcare. Through its use, surgical job automation appears to have made significant progress, which will ultimately benefit patient outcomes. Adding 3D video data for increased precision, extending the system's capacity to manage more complicated surgeries, and utilising cutting-edge AI methods like reinforcement learning to continuously improve surgical support models are some examples of future improvements.

References

1. Hussain, S.M.: Explainable deep learning for medical image processing: computer-aided diagnosis and robot-assisted surgery (2023)
2. Albahri, A.S., Duhaim, A.M., Fadhel, M.A., Alnoor, A., Baqer, N.S., Alzubaidi, L., Deveci, M.: A systematic review of trustworthy and explainable artificial intelligence in healthcare: assessment of quality, bias risk, and data fusion. Inf. Fusion **96**, 156–191 (2023)
3. Teixeira, K., Miguel, G., Silva, H.S., Madeiro, F.: A survey on applications of unmanned aerial vehicles using machine learning. IEEE Access (2023)
4. Kok, E.N., Eppenga, R., Kuhlmann, K.F., Groen, H.C., van Veen, R., van Dieren, J.M., Ruers, T.J.: Accurate surgical navigation with real-time tumor tracking in cancer surgery. NPJ Precis. Oncol. **4**(1), 8 (2020)
5. Mendizabal, A.: Machine learning meets real-time numerical simulation-application to surgical training, preoperative planning, and surgical assistance. Doctoral dissertation, Université de Strasbourg (2020)
6. Hussain, S.M., Brunetti, A., Lucarelli, G., Memeo, R., Bevilacqua, V., Buongiorno, D.: Deep learning-based image processing for robot-assisted surgery: a systematic literature survey. IEEE Access **10**, 122627–122657 (2022)
7. Aboutahoun, D., Zewail, R., Soliman, M.I.: On potentials of few-shot learning for AI-enabled Internet of Medical Things. In: 2022 IEEE Globecom Workshops (GC Wkshps), pp. 1062–1067 (2022)
8. Bonda, D.J., Pruitt, R., Goldstein, T., Varghese, A., Shah, A., Rodgers, S.: Robotic surgical assistant rehearsal: Combining 3-dimensional-printing technology with preoperative stereotactic planning for placement of stereoencephalography electrodes. Oper. Neurosurgery **19**(2), 190–194 (2020)
9. Wang, M., Li, D., Shang, X., Wang, J.: A review of computer-assisted orthopaedic surgery systems. Int. J. Med. Robot. Comput. Assist. Surgery **16**(5), 1–28 (2020)
10. Jacobsen, M.F., Konge, L., Alberti, M., La Cour, M., Park, Y.S., Thomsen, A.S.S.: Robot-assisted vitreoretinal surgery improves surgical accuracy compared with manual surgery: a randomized trial in a simulated setting. Retina **40**(11), 2091–2098 (2020)
11. Prevost, G.A., Eigl, B., Paolucci, I., Rudolph, T., Peterhans, M., Weber, S., Lachenmayer, A.: Efficiency, accuracy and clinical applicability of a new image-guided surgery system in 3D laparoscopic liver surgery. J. Gastrointestinal Surgery **24**(10), 2251–2258 (2020)

12. Lonner, J.H., Seidenstein, A.D., Charters, M.A., North, W.T., Cafferky, N.L., Durbhakula, S.M., Kamath, A.F.: Improved accuracy and reproducibility of a novel CT-free robotic surgical assistant for medial unicompartmental knee arthroplasty compared to conventional instrumentation: a cadaveric study. Knee Surg. Sports Traumatol. Arthrosc. **30**(8), 2759–2767 (2022)
13. Mathkor, D.M., Mathkor, N., Bassfar, Z., Bantun, F., Slama, P., Ahmad, F., Haque, S.: Multirole of the Internet of Medical Things (IoMT) in biomedical systems for managing smart healthcare systems: an overview of current and future innovative trends. J. Infect. Public Health (2024)
14. Sharma, K.: Personalized telemedicine utilizing artificial intelligence, robotics, and Internet of Medical Things (IoMT). In: Handbook of Research on Artificial Intelligence and Soft Computing Techniques in Personalized Healthcare Services, pp. 301–323. Apple Academic Press (2024)
15. Pradyumna, G.R., Hegde, R.B., Bommegowda, K.B., Jan, T., Naik, G.R.: Empowering healthcare with IoMT: evolution, machine learning integration, security, and interoperability challenges. IEEE Access (2024)
16. Baseer, K.K., Sivakumar, K., Veeraiah, D., Chhabra, G., Lakineni, P.K., Pasha, M.J., et al.: Healthcare diagnostics with an adaptive deep learning model integrated with the Internet of Medical Things (IoMT) for predicting heart disease. Biomed. Signal Process. Control **92**, 105988 (2024)
17. Khan, A., Rizwan, M., Bagdasar, O., Alabdulatif, A., Alamro, S., Alnajim, A.: Deep learning-driven anomaly detection for IoMT-based smart healthcare systems. CMES-Comp. Model. Eng. Sci. **141**(3) (2024)
18. Almujally, N.A., Aljrees, T., Umer, M., Saidani, O., Hanif, D., Abuzinadah, N., et al.: IoMT based smart healthcare system to control outbreaks of the COVID-19 pandemic. PeerJ Comp. Sci. **9**, e1493 (2023)
19. Zhang, Y.: Enhancing surgical gesture recognition using Bidirectional LSTM and evolutionary computation: a machine learning approach to improving robotic-assisted surgery. Doctoral dissertation (2024)
20. Alshirbaji, T.A., Jalal, N.A., Docherty, P.D., Neumuth, T., Möller, K.: A deep learning spatial-temporal framework for detecting surgical tools in laparoscopic videos. Biomed. Signal Process. Control **68**, 102801 (2021)
21. Nwoye, C.I.: Deep learning methods for the detection and recognition of surgical tools and activities in laparoscopic videos. Doctoral dissertation, Université de Strasbourg (2021)
22. Luongo, F., Hakim, R., Nguyen, J.H., Anandkumar, A., Hung, A.J.: Deep learning-based computer vision to recognize and classify suturing gestures in robot-assisted surgery. Surgery **169**(5), 1240–1244 (2021)
23. Xu, J., Shen, J., Wan, C., Jiang, Q., Yan, Z., Yang, W.: A few-shot learning-based retinal vessel segmentation method for assisting in the central serous chorioretinopathy laser surgery. Front. Med. **9**, 821565 (2022)
24. Killeen, B.D., Chaudhary, S., Osgood, G., Unberath, M.: Take a shot! Natural language control of intelligent robotic X-ray systems in surgery. Int. J. Comp. Assis. Radiol. Surgery, 1–9 (2024)
25. Sadraoui, A., Martin, S., Barbot, E., Laurent-Bellue, A., Pesquet, J.C., Guettier, C., Ayed, I.B.: A transductive few-shot learning approach for classification of digital histopathological slides from liver cancer. In: 2024 IEEE International Symposium on Biomedical Imaging (ISBI), pp. 1–5. IEEE (2024)
26. Nazari, J., Motie Nasrabadi, A., Menhaj, M.B., Raiesdana, S.: Epilepsy seizure prediction with few-shot learning method. Brain Inf. **9**(1), 21 (2022)
27. Han, Z., Tian, H., Han, X., Wu, J., Zhang, W., Li, C., et al.: A respiratory motion prediction method based on LSTM-AE with attention mechanism for spine surgery. Cyborg Bionic Syst. **5**, 0063 (2024)
28. Sitaraman, S.R., Alagarsundaram, P., Thanjaivadivel, M.: AI-driven robotic automation and IoMT-based chronic kidney disease prediction utilizing attention-based LSTM and ANFIS. Int. J. Multidiscip. Educ. Res. **13**(8) (2024)
29. Sitaraman, S.R.: Advanced IoMT-enabled chronic kidney disease prediction leveraging robotic automation with autoencoder-LSTM and fuzzy cognitive maps. Int. J. Mod. Electron. Commun. Eng. **12**(3) (2024)

30. Alagarsundaram, P., Sitaraman, S.R., Gollavilli, V.S., Gattupalli, K., Nagarajan, H., Adewole, K.S.: Adaptive CNN-LSTM and neuro-fuzzy integration for edge AI and IoMT-enabled chronic kidney disease prediction. Int. J. Appl. Sci. Eng. Manage. **18**(3) (2024)
31. https://www.kaggle.com/code/yannicksteph/u-net-cv-brain-prediction/input

Integration of Wearable Devices in Healthcare: The Need to Examine Their Implications for Health Workers

Virginia Gunn, Philip Eappen, Hikmat Singh Brar, Emma Brulin, and Carles Muntaner

Abstract Despite a continuous increase in the use of wearable devices in healthcare to support remote health monitoring, knowledge about the implications of these technologies for health workers is scarce. Given that health workers are among the most valuable and scarce resources within health systems, protecting their health, safety, and well-being should be a major priority. It is therefore important to have a thorough understanding of the ways in which health workers may be impacted by the increased use of wearable devise technologies. In turn, this information could be used to prevent or mitigate possible negative implications for workers' working conditions, health, and well-being. In this chapter, we provide a brief overview of wearable devices used in healthcare, distinguishing between those used by patients and those used by health providers. We also review potential implications of the use of these devices for health workers' employment and working conditions and formulate several research, practice, and policy recommendations.

Keywords Increased monitoring of health metrics · Patient care · Technology adoption · Healthcare delivery · Advanced health technologies · Occupational health and safety · Health system sustainability

V. Gunn (✉) · P. Eappen · H. S. Brar
Cape Breton University, Sydney, NS, Canada
e-mail: virginia_gunn@cbu.ca

V. Gunn · E. Brulin
Unit of Occupational Medicine, Institute of Environmental Medicine, Karolinska Institute, Solna, Sweden

C. Muntaner
Lawrences F. Bloomberg Faculty of Nursing and Dalla Lana School of Public Health, University of Toronto, Toronto, Canada

P. Eappen et al. (eds.), *Remote Monitoring and Wearable Devices in Healthcare*, Information Systems Engineering and Management 63,
https://doi.org/10.1007/978-3-031-98897-4_6

1 Introduction

In recent decades, the healthcare field has bared witness to an accelerated increase in health technologies with potential to reshape healthcare delivery [1, 2], including wearable devices (e.g., smart sensors, electronic wearables), a particular type of health technologies, consisting of devices worn either by patients [3–5] or by health workers [6–9]. As described in depth in this book, and briefly summarized in this chapter, there are numerous types of devices used across a wide range of health-related contexts and health institutions.

Wearable devices serve diverse purposes, although overall, they seem to fit four key aims: (i) increased monitoring of health metrics to support improvement of health outcomes [10–14], (ii) reduced use of human and material resources and increased health institution efficiencies [15, 16], (iii) increased monitoring of health workers' activities or interactions to improve efficiencies or other aspects of care provision [8], and (iv) improved occupational health and safety for health workers [6, 7]. As expected with any new technologies, the use of wearable devices within the healthcare field will change working methods and the organization of work, with numerous possible, intended or unintended, implications for health workers' employment and working conditions. For example, the use of wearable devices, along with other health technologies, could result in changes to health workers employment security, pay rates, physical and psychosocial working conditions [17], education, or training expectations [18].

Protecting worker health and well-being should be a vital concern for health institutions given that workers constitute one of the most valuable and scarce assets they have, as acutely reminded during the COVID-19 pandemic that placed increased pressures on health systems around the world [19]. However, despite a shared recognition that ensuring the sustainability of the health workforce should be a priority [19], protecting the health and well-being of health workers is not usually among the most targeted strategies aimed at safeguarding the workforce. To this point, despite a rapid increase in the use of health technologies in healthcare [1, 2], including wearable devices, most research is concerned mainly with examining the use of such devices and their implications for patient outcomes and institutional costs and efficiencies [15, 20]. Meanwhile, examining the impact of advanced health technologies such as artificial intelligence on workers' employment and working conditions, and their health and well-being is often neglected or pursued inconsistently [17]. Accordingly, we strongly believe that, in spite of the worthy goals fueling the use of wearable devices (e.g., improved patient health outcomes and increased health system efficiencies), overlooking the possible implications of such devices for health workers, including the possible negative effects, will endanger the sustainability of the health workforce and that of the health systems they are meant to preserve. In this chapter, in addition to discussing several key links between wearable devices and health workers employment and working conditions, we make several research, practice, and policy recommendations to advance work on this topic and address important knowledge gaps.

2 Overview of Wearable Devices Used in Healthcare

2.1 *Wearable Devices Designed for Patient Use*

Currently, most wearable devices used within the healthcare system are designed for patient use. Thus, a wide range of miniature digital sensors incorporated into watches, bands, glasses, or monitors are worn by patients from the comfort of their home or during their visits or stays in health facilities. The monitoring of health status indicators collected through the use of digital devices is shared between patients and health workers, while the interpretation of findings is typically within the realm of health workers. Generally, wearable devices enable access to consistent and accurate recordings of health indicators, providing health workers with additional wide-ranging data to make informed healthcare decisions for their patients [4, 21]. Thus, real time tracking of various physiological markers such as body temperature, sleep health, blood glucose, blood pressure, heart rate, or heart activity, allows health workers to 'keep an eye' on patients and respond promptly to emergencies [22, 23]. As expected, wearables range from simple, widely used, and easily accessible tools meant to support well-being (e.g., smartwatches, fitness trackers, and glucose sensors) [4, 21] to more complex, cutting-edge technologies that can provide ongoing, advanced, and cost-effective health monitoring (e.g., Wireless Body Sensor Networks (WBSNs), or printed smart devices) [20, 24].

Trailblazing technologies such as artificial intelligence (AI) further increase the potential of wearable devices to support health status monitoring. For instance, the internet of things (IoT) eases the integration and sharing of data and thus, provides caregivers with convenient access to patient data through smartphones and cloud-based systems [22, 25]. Additionally, AI-enabled sensors and IoT frameworks considerably expand the possible applications of wearable devices and facilitate their incorporation into telemedicine digital health care approaches enabling, for instance, the provision of care via distance, real time analysis, contact tracing, and prediction of disease spreading during pandemics [21, 26, 27].

Unsurprisingly, the potential for patients to benefit from the use of wearable devices is high, especially for patients with access to health systems with advanced infrastructures and adequate integration of wearable devices within hospital digital systems [28] and patients who live in areas with reliable internet connectivity. Key patient advantages include (i) real-time feedback on lifestyle behaviors, which is conducive to improved health behaviors and increased engagement in care activities and ownership [12, 21, 29, 30], (ii) improved access to health care [14, 31], (iii) earlier identification of health issues and relevant clinical interventions, and better outcomes [5, 23, 32, 33], along with (iv) remote symptom management, and (v) reduced hospital and emergency visits [10, 13, 14, 31].

2.2 Wearable Devices Designed for Health Worker Use

Gradually, the largely patient focus of wearable devices has started to incorporate a health worker focus. Thus, several wearable devices are now used by health workers including, for example, bodily sensors, proximity beacons, smart biomedical devices, or robotics systems, all designed to help monitor workers' physical interactions, thereby ensuring that safety protocols related to physical distancing are followed, especially during pandemics [34, 35]. Such devices reduce the risk of acquired infections for health workers by making it easier to ensure physical distancing or by supporting remote treatments [34, 36]. Another example of wearables used by workers consists of devices used to measure short- or long-term exposures to occupational health hazards and issue alerts when safe levels are exceeded [7, 17], devices used to monitor body posture, safe lifting techniques, or ergonomics [7, 37], or health and well-being [38, 39] during work activities.

2.3 Applicability of Wearable Devices Across Health Specialties and Settings

Wearable devices meant to improve disease management through increased monitoring of health metrics are used across a wide variety of health settings including (i) primary and community care, (ii) acute and urgent care, (iii) long-term/elderly/palliative care, as exemplified next through several examples.

Primary and community care. For instance, within primary and community care, wearables are used to monitor and facilitate self-management of diabetes, cardiovascular diseases, respiratory illnesses, mental health issues, or epilepsy [4, 10, 26, 40]. Smartphones and wearable devices used for activity recognition and classification support health monitoring and assistance for the elderly in community settings [41], while IoT-enabled wearables enhance remote patient management by transmitting real-time data to caregivers to ensure continuity of care outside clinical facilities [42, 43].

Acute and urgent care. In acute care, skin-wearable sensors offer non-invasive monitoring of thermal, neural, and photoelectrical signals for immediate diagnostic and intervention [42]. Wearable devices in intensive care units track mobility, gait, and physical recovery post-surgery, relying on accelerometers to assess outcomes such as step count or walking speed [44]. Medical-grade wearables for monitoring critical health parameters like electrocardiogram activity, oxygen saturation, and blood pressure during surgical recovery have been used for a long time [4]. Further, advanced wearable systems using sensor-based triage and telemedicine technologies streamline patient monitoring during high-pressure situations in hospitals [43].

Long-term/elderly/palliative care. Wearable devices in long-term and palliative care allow continuous symptom monitoring, management, and treatment adjustments for older adults or terminal patients, reducing the need for clinic visits while maintaining communication with health workers [13]. Devices monitoring heart rate variability and sleep patterns provide insights into mental health for long-term care patients and their caregivers [10]. Although the actual precision needed for elderly patients is not currently reflected across all devices [11, 45] and these technologies demand improved infrastructure and a highly trained workforce [18, 31], sensor-based AI technologies have potential to enhance independent living and provide useful insights to caregivers regarding patients' health, thus enabling better decision-making [31]. Further, wearable accelerometers, actigraph units, and other sensors track patient mobility, human rest/activity cycles, and monitor health status in palliative care, enabling tailored treatments and symptom management [13]. Additionally, wearable devices enable patient-centered care, allowing caregivers to adjust treatments without requiring frequent clinic visits, a critical factor for elderly and terminal patients [13].

Given that many wearable devices allow health workers to monitor patients remotely, all health settings and areas that use such devices could benefit from advantages such as reduced overall healthcare expenditure despite an increase in care provision and accessibility [14, 31]. Of course, the upfront investments in technology should be considered, which means that the savings are not immediate. Another advantage is derived from the ability of health workers to handle a higher patient load, whose assessment data they could analyse in real time, to more quickly diagnose and manage various acute and chronic conditions such as diabetes, sleep disorders, or cardiac diseases [26, 32]. Additionally, the use of wearable devises by patients may further decrease health institution costs through reduced hospital visits, stays, and re-admissions [10, 14, 31].

However, while it is obvious that all potential advantages mentioned in this section may have significant appeal for health institutions, it is vital to study the actual impact of wearable devices on patient care outcomes and financial efficiencies. Equally important is to examine closely the impact of wearable devices on the employment and working conditions of workers and their health and well-being.

3 Possible Implications of the Use of Wearable Devices in Healthcare for Health Workers

3.1 Job Satisfaction, Well-Being, and Occupational Health and Safety Implications

Numerous wearable devices have been eagerly adopted in healthcare as a solution to increased care demands. Such devices are praised for their role in improved monitoring, communication, and maintaining the overall health of both patients and health workers. As described next, although wearable devices may ease workload and stress for health workers and improve their work satisfaction, because they introduce new models of care and changes in the organization of work, they may also create increased job demands and strain.

Wearable devices may lead to a more efficient and effective utilization of health workers. For example, because monitoring devices keep track of routine patients' signs and symptoms, they free up time for health workers to focus on patients with high acuity who require in person monitoring, thus possibly easing their current overloads [15, 46]. Further, remote patient monitoring through wearable devices may assist health workers with the completion of tasks, including automated real-time health and safety monitoring and data sharing [4, 38, 47], chronic disease management, disease diagnosis, treatment, and rehabilitation [4]. In turn, enhanced monitoring has potential to improve care quality while minimizing stress on acute services like ambulance and emergency rooms [16, 46] or primary care facilities, thereby, cutting down in-person visits and the overall patient burden on health workers and health facilities.

Additionally, wearable devices improve communication between patients and health workers by enabling real-time updates, a clear advantage when compared to previous ways of sharing information (e.g., health status updates) with providers, typically taking place only when patients had appointments with health workers. In turn, improved information exchanges help facilitate quick and timely care interventions to address medical issues [16] and thus, improve health outcomes, which, unsurprisingly, contributes to increased patient trust in care providers. Wearable tools may also improve care quality in other ways by (i) promoting multidisciplinary approaches to patient care, (ii) enhancing communications between the health facilities involved in the provision of patient care [10, 46], or reducing isolation or other potential shortcomings of home care for discharged patients [28]. Additionally, linking wearable devises to patient e-records can assist in making more coordinated and informed care decisions.

Unsurprisingly, improving health status monitoring and the quality and safety of patient care may, in turn, contribute to a sense of meaningfulness, well-being, and job satisfaction for health workers [48]. Thus, the use of wearable devices and remote patient monitoring systems is linked to an increased number of health workers being satisfied at work [49]. Positive experiences with the seamless use of such devices and the tools' ability to reduce workload and the need for health workers' in-person presence for care provision has potential to improve work efficiency, job satisfaction, and staff morale [15, 16]. In turn, increased job satisfaction and morale have potential to help workers develop a feeling of control, building their confidence and supporting a safety culture and positive work environment where workers feel valued and supported.

At the same time, the use of wearable devices in the provision of care might contribute to increased job demands, new learning needs, and technostress [49]. Technostress is found to negatively impact productivity, job performance, job satisfaction and organizational commitment [49, 50] as related to constant changes and upgrades of software and hardware, ongoing equipment alarms and need for troubleshooting, and the fear of becoming obsolete at work. Moreover, the wide application of wearable devices might result in a large amount of information collected, even information that is not needed in its entirety, which requires careful storage and handling once collected. Additionally, the increased use of wearables may also lead to job alienation and work-life balance challenges due to constant monitoring expectations [51]. Further, unreliable devices, technological uncertainty, frequent interference from system updates, or sudden technical changes might lead to moral distress among health workers because they believe wearable devices are preventing them from delivering the high-quality patient care they aim for [50].

Wearable devices may also be used to promote occupational health and safety among health workers and prevent job hazards. For instance, various wearable tools are used to (i) identify and monitor fatigue, exposure to hazards, or stress [7, 38, 52]; (ii) report on air quality, noise, and other job exposures [6]; (iii) monitor sleep deprivation, physical activity, and heart rate to analyze signs of stress and burnout so that it can be diagnosed and managed as early as possible [38, 39]; (v) support ergonomic risk assessments [9]; and (vi) reduce physical strain and repetitive tasks, especially in settings like nursing and occupational health [7, 52], thereby increasing workplace safety for health workers. Despite their positive role towards improving occupational health and safety, the wearing of technical devices while working has raised questions about the possible physical strain they cause, including the impeding of workers' movement.

3.2 *Implications for Workers' Employment and Working Conditions*

As emphasized earlier, wearable devices are being increasingly integrated into healthcare settings, offering both potential benefits and challenges for health workers [53, 54]. Yet their effect on workers' employment and working conditions has attracted less attention in the literature. Regarding employment and working conditions, wearable devices have the potential to enhance work efficiency, improve physical well-being, and reduce work-related injuries. Despite these benefits, there are several technology-related, social, and economic challenges that require addressing to increase the smooth and ethical adoption of wearables in healthcare, as briefly introduced in this chapter.

The use of wearable technologies is prompting a significant overhaul in the delivery of healthcare, reshaping work organization in the healthcare industry by changing job design and staff responsibilities. Such technologies affect job autonomy, skill requirements, and performance feedback systems which, in turn, require new work organization frameworks and legislation [17, 53–55]. The wearable devices relying on cutting-edge technologies require enhanced skill sets to operate, manage, and interpret their data outputs. While wearables help with automating repetitive tasks and optimizing work processes, they could limit health worker creativity and problem-solving skills through work standardization or routinization [55]. With a constant flux of upcoming new technology and devices, health workers are also expected to maintain their proficiency in terms of skills needed to operate these new tools, which reinforces the need for ongoing workforce training to maintain relevance in the healthcare field [7, 18, 56]. The ongoing shift to a widespread use of digital wearable devices in healthcare underscores the rapid requirement to make health professional status compatible with the use of technical advancements. Thus, health workers should be well versed with the new advancements and features of such devices to ensure they are able to put wearables to maximum use while not losing track of their potential ethical implications for patient care [56]. Appropriate training and practice is needed to ensure that health workers are confident while using these devices, as they must focus on multiple aspects including device usage, integration, and data privacy [7]. Adequately trained health workers will make the use of wearables easy and beneficial while also minimizing the chances of error, data loss, or privacy breaches.

The continuous incorporation of wearable devices will bring upon a massive shift in the work culture of health workers, presenting potential to create a more efficient and satisfied workforce. It has the potential to redefine job roles, educational requirements, and clinical training. Nevertheless, the adoption of wearables in clinical practice is hindered by challenges such as security concerns, lack of industry standards, user-friendliness [4], or compromised worker rights [53]. These devices pose serious concerns with respect to ethics, consent, and data privacy, raising doubts about the sufficiency of current regulatory structures and healthcare institutions' ability to maintain data privacy and security [54, 57–62]. Concerns about data privacy and security are not limited to wearable devices but are commonly shared across many new technologies adopted in healthcare in recent years, including machine learning and the IoT applications [2, 58, 63]. Such concerns serve as reminders for the need for safety measures to keep classified patient and worker information protected and compliant with relevant regulations [7, 52], especially since shortcomings of current regulatory structures put both employers and workers in vulnerable positions [53].

Wearable technologies have potential to increase positive interactions between health workers and patients and enable smooth patient experiences, thus, providing worker motivation and a sense of engagement because of their capacity to simplify work and improve job satisfaction [17]. However, a dependence on technology to gather health metrics comes at the cost of losing professional human touch, clinical assessment skills, and empathy, which is concerning, and reinforces calls for wearable devices to be used to complement rather than replace human expertise [54]. Research on hygiene practices in health facilities illustrates this point. For example, a study exploring the impact of a novel electronic wearable device on hand hygiene compliance among health workers found that while the device itself did not significantly improve compliance with international guidelines, the device, along with training and worker motivation, successfully enhanced the volume and duration of hand hygiene actions [8].

As already discussed at length in this chapter, the use of wearable devices in healthcare significantly enhances the capabilities of front-line health workers by providing them with access to real-time data that supports chronic disease management, rehabilitation, and safety monitoring [52]. These technologies foster teamwork and collaboration among health workers, facilitating shared insights and workflow efficiencies, all of which have potential to improve patient care and health outcomes [10, 47, 64]. However, an increased reliance on wearables also adds to the responsibilities of health workers, requiring them to manage extensive data sets, monitor remote patients and address their queries, which can, in turn, lead to new workloads and stress [65]. Furthermore, the risk of self-diagnosis associated with wearable technologies could disrupt the traditional health worker-patient relationship, complicating clinical interactions and decision-making processes [65]. As health workers navigate these challenges, it is essential to assess the broader implications of wearable technology on their roles and well-being. Such assessments would help ensure that the benefits of enhanced monitoring are achieved without compromising the professional effectiveness or mental health of health workers [52, 66].

In summary, with the dawn of advanced wearable devices and AI technologies in healthcare, we witness a significant transformation in work culture, with changes in employment and working conditions, including work autonomy, occupational safety, and training needs along with various additional ethical and legal concerns. While wearable devices offer significant potential to improve healthcare delivery and worker well-being, their integration into healthcare settings requires overcoming several challenges, including improving device accuracy, ensuring data privacy, and establishing regulatory standards [57] along with increasing understanding of the impact of using such devices on health workers, as captured next in more detail in the recommendations section.

4 Recommendations for Research, Practice, and Policy

4.1 Research Recommendations

While the use of wearable devices in healthcare shows potential to improve health outcomes and employment and working conditions for health workers, empirical evidence is required to substantiate these claims. Thus, researchers should conduct both quantitative (e.g., longitudinal and cross-sectional) and qualitative studies to examine the impact of wearable devices on health workers' health and safety, and employment and working conditions, and establish and collect relevant indicators. Further, to ensure high adoption rates, wearable devices designed for health worker use must be user-friendly and comfortable to wear for extended periods [4, 67], as well as adherent to standards of infection control. Research should focus on testing ergonomic designs and user interface improvements to enhance health workers experiences and encourage consistent use. When such technologies are worn by health workers, especially for long periods during their shifts, there is a risk that they are exposed to electromagnetic fields [68]. Accordingly, researchers should develop guidelines and risk assessment protocols to better understand the risks posed by these exposures to health workers' health and mitigate them to ensure the safe use of wearable devices in healthcare [68].

Additionally, both the trustworthiness and accuracy of sensors used in wearable devices are important for ensuring that health workers receive accurate health monitoring data that supports rather than impede or slow down their efforts to provide effective care. Thus, future research should aim to improve sensor technology to enhance the precision of data collected, particularly in high-stakes environments like hospitals where accurate health monitoring is essential [69]. An important consideration with wearable devices is data privacy and security. Given that many of these devices gather sensitive patient data, it is critical to develop strong security protocols to protect this information. Research should focus on creating secure data transmission and storage solutions to prevent unauthorized access and breaches [4] that could negatively affect patients and health workers alike. Equally important, research should examine methods for integrating wearable data with various healthcare management and documentation systems including electronic health records (EHRs); this approach would enable health workers to capture a thorough view of a patients' health status and incorporate it into existing documentation systems [70], and thus, streamline their work and increase documentation efficiencies. Further, the cost of implementing wearable technologies in healthcare settings could be a barrier to widespread adoption across health setting and patient populations [71], which could cause ethical dilemmas for health workers since they may not be able to use them consistently with all of their patients. Research should study cost-effective solutions and models that can increase the widespread adoption of these technologies by a broader range of healthcare facilities and patient populations.

While the use of wearable devices in healthcare settings represents a great opportunity for supporting the provision of care and strengthening the health and safety of health workers, further research is needed to fill existing knowledge gaps and ensure these technologies are utilized to their best. Such research should focus on the impacts of wearable devices in healthcare for worker health, well-being, and employment and working conditions along with related implications for costs, design, reliability, and health system integration. Addressing these areas will help maximize the benefits of wearable technology in healthcare settings and for health workers.

4.2 Practice Recommendations

Developing and integrating new solutions, or work methods, such as wearable devices, in healthcare should aim to equally benefit patients, health workers, and healthcare organizations [72–74], as briefly overviewed next. To ensure equal benefit, both the development and implementation of wearable devices in healthcare should adopt the systemic perspective of Healthy Healthcare (HHC) [72–74]. HHC acknowledges the complex and interdependent relationship between these three main pillars in healthcare: (i) health workers, (ii) organizational practice, and (iii) quality and safety of patient care. By involving the three pillars, wearable devises may be developed and implemented to meet the needs of health workers (providing support in their work and easing job demands), patients (providing access to safe and quality care) and organizations (increasing effectiveness and efficiency), thus, increasing their likelihood of being successful in both the short- and long-term.

Further, the implementation of wearable devices should be guided by evidence-informed risk–benefit analysis results. While wearable devices may lead to more effective and efficient forms of care or treatment, they could also result in additional and increasing healthcare costs in the short-term and may be followed, in some cases, by detrimental outcomes for staff well-being and the organization of services if the human factor (health workers in this case) is not sufficiently accounted for. To ensure the successful introduction of wearable devices in healthcare, the development of these tools, their implementation, and evaluation should therefore adopt an HHC perspective. The use and integration of wearables into healthcare is promising in terms of worker support and patient outcomes. Adopting an HHC perspective while identifying the practical areas of need for application of these wearables has potential to increase the likelihood of improving these tools' usability and effectiveness.

Wearable devices are increasingly recognized for their potential to enhance out-of-hospital care, and reduce healthcare expenditures by improving efficiencies and streamlining patient management [75, 76]. However, the successful implementation of these technologies into existing healthcare models requires careful consideration of reimbursement frameworks, user-friendly designs, and tailored solutions that address the diverse needs of both patients and health workers [75], along with assessments to determine whether the HHC perspective is an adequate tool. Equally important is the acknowledgement that wearable devices demand technical knowledge and skills and that potential changes to work tasks and workload should be assessed. Both healthcare workers and patients need the training necessary for accurate interpretation, use, and integration of data collected through the use of wearables into practice [30]. Adequate training for both front-line workers and management is essential to fully harness the capabilities of wearable technologies in clinical settings [77]. Such training should also emphasize the practical application of wearables, ensuring that staff can effectively interpret and utilize data to inform patient care decisions [77]. Additionally, addressing potential barriers—such as privacy concerns and the need for intuitive interfaces—is crucial to fostering acceptance and maximizing the benefits of wearable health monitoring technologies [52, 78].

While wearables hold promise for transforming healthcare delivery, their development and implementation must be strategic and inclusive, prioritizing the needs of all actors [75], including health workers, organizations and management, and patients, who must feel confident and motivated to engage with wearable technologies. By addressing these multifaceted challenges, the healthcare system can harness the full potential of wearable devices to improve patient outcomes and reduce costs effectively.

4.3 Policy Recommendations

Given that wearable devices could impact health workers across several dimensions, particularly with regard to occupational health and safety, and working conditions (e.g., professional autonomy, skill requirement, education and clinical training needs, or job satisfaction) [58, 79], strong and comprehensive legislation and policies are essential to protect workers from the possible risks posed by such technologies [53]. It is therefore important to adopt policies that address possible negative outcomes such as technostress, physical strain, hygiene, or radiation concerns from prolonged use of devices [68, 80] affecting health workers using wearable devices. Moreover, given that integrating AI-powered wearables can lead to over-reliance on technology, potentially diminishing clinicians' critical thinking and decision-making autonomy [81, 82], there is a need for policies that protect workers' professional autonomy and call for their meaningful involvement in the design of technologies used in their job. Given a range of ethical and privacy concerns emerging from the use of wearable devices, particularly regarding data security and the ownership of health data [83, 84], there is a need for policies addressing these concerns.

Further, the advancement of technologies and wearable devices in healthcare necessitates a careful balance between innovation and the protection of workers' rights, reflecting European regulations such as the General Data Protection Regulation and the AI Act. These frameworks are essential for ensuring that the implementation of wearable technologies is both legitimate and respectful of employee autonomy and data privacy [59]. Regulatory bodies must establish guidelines that allow for the flexible use of wearables while safeguarding personal data and ensuring compliance with ISO standards for privacy and security [60]. Moreover, health workers must be well-aware of local regulations to mitigate liability risks associated with using wearables in clinical settings [61]. Ethical considerations, particularly regarding equity and fairness in digital health, should be reflected in the design of policy and legislation guiding the use of wearable devices [62]. For instance, the potential for dehumanization in healthcare, driven by over-reliance on technology, underscores the need for regulations that reinforce ethical standards in deploying wearable technologies [62]. Overall, a comprehensive policy approach that integrates legal, moral, and practical considerations is vital for advancing technologies in healthcare.

A minimum set of policy and legislative recommendations is essential in every country to harness the potential of wearable technologies in healthcare while safeguarding user rights and promoting equitable access. Regulatory frameworks should be established to ensure that wearable devices are designed with user privacy, rights, and data protection at the core, in alignment with existing laws [54, 57–62]. Laws should include comprehensive guidelines on data handling and user consent [85, 86]. Moreover, addressing barriers to adoption, such as affordability and accessibility, is crucial for promoting health equity. For instance, policy initiatives could be implemented to subsidize wearable devices for underserved populations [87]. Additionally, training programs for health workers on the ethical use of wearables and their implications for patient autonomy should be mandated, to ensure practitioners are able to navigate the complexities of wearable technologies in clinical settings [3]. Finally, ongoing research and involved actor engagement are vital to continuously refine policies based on user experiences and emerging challenges to mitigate the risks associated with wearable technologies while maximizing their benefits for patients and health workers [88].

5 Conclusion

In this chapter, we summarize several categories of wearable devices used by patients across a wide range of health-related contexts and health institutions. Although very diverse, such devices serve four key aims: (i) increased monitoring of health metrics to support improvement of health outcomes, (ii) reduced use of human and material resources and increased health institution efficiencies, (iii) increased monitoring of health workers' activities or interactions to improve efficiencies or other aspects of care provision, and (iv) improved occupational health and safety for health workers. We also review several wearable devices worn by health workers and the overall purposes they fit, including (i) monitoring worker well-being or exposure to occupational health hazards, (ii) supporting ergonomic risk assessments related to body posture, or safe lifting techniques, or (iii) determining physical proximity to patients or other workers.

Despite the rapid and continuously increasing use of wearable devices in health care, most research examines the use of such devices and their implications mainly from the perspective of patient outcomes, or institutional costs and efficiencies. Meanwhile, examining the impact of health technologies such as wearable devices or AI-based applications on workers' employment and working conditions, and their health and well-being is often neglected or pursued inconsistently. As expected with any new technologies, the use of wearable devices within the healthcare field will change working methods and the organization of work, with numerous possible, intended or unintended, implications for health workers' employment and working conditions, as discussed in this chapter. Given that protecting worker health and well-being should be a vital concern for health institutions since workers constitute one of the most valuable and scarce assets they have, we advocate for increased attention to this aspect and make several research, practice, and policy recommendations to advance work on this topic and address important knowledge gaps.

References

1. Eappen, P., Vajjhala, N.R.: Healthcare informatics innovation post-Covid-19: lessons, challenges, and opportunities. In: Eappen, P., Vajjhala, N.R. (eds.) Healthcare Informatics Innovation Post Covid-19 Pandemic, 1st edn. Routledge, Taylor and Francis & CRC Press (2025)

2. Vajjhala, N.R., Eappen, P.: Smart health: advancements in machine learning and the internet of things solutions. In: Samui, P., Sekhar Roy, S., Zhang, W., Taguchi, Y.H. (eds.) Machine Learning and IoT Applications for Health Informatics, pp. 31–51. CRC Press (2024)
3. Dunn, J., Runge, R., Snyder, M.: Wearables and the medical revolution. Pers. Med. **15**(5), 429–448 (2018)
4. Lu, L., Zhang, J., Xie, Y., Gao, F., Xu, S., Wu, X., et al.: Wearable health devices in health care: narrative systematic review. JMIR Mhealth Uhealth **8**(11), e18907 (2020). https://doi.org/10.2196/18907
5. Mukherjee, S., Suleman, S., Pilloton, R., Narang, J., Rani, K.: State of the art in smart portable, wearable, ingestible and implantable devices for health status monitoring and disease management. Sensors. **22**(11), 4228 (2022)
6. Mueller, W., Smith, A., Kuijpers, E., Pronk, A., Loh, M.: Worker perspectives on improving occupational health and safety using wearable sensors: a cross-sectional survey. Ann. Work. Expo. Health **68**(8), 867–873 (2024)
7. Patel, V., Chesmore, A., Legner, C.M., Pandey, S.: Trends in workplace wearable technologies and connected-worker solutions for next-generation occupational safety, health, and productivity. Adv. Intell. Syst. **4**(1), 2100099 (2022). https://doi.org/10.1002/aisy.202100099
8. Pires, D., Gayet-Ageron, A., Guitart, C., Robert, Y.-A., Fankhauser, C., Tartari, E., et al.: Effect of wearing a novel electronic wearable device on hand hygiene compliance among health care workers: a stepped-wedge cluster randomized clinical trial. JAMA Netw. Open **4**(2), e2035331 (2021). https://doi.org/10.1001/jamanetworkopen.2020.35331.
9. Sabino, I., do Carmo Fernandes, M., Cepeda, C., Quaresma, C., Gamboa, H., Nunes, I.L., et al.: Application of wearable technology for the ergonomic risk assessment of healthcare professionals: a systematic literature review. Int. J. Ind. Ergon. **100**, 103570 (2024)
10. Hassan Almansour, H.A., Almanajam, F.A., Alyami, M.A., Shawil Al-Mahamad, A.H., Saleh Al Zabid, N.H., Hamad Al-Mahamad, M.S., et al.: Evaluating the impact of wearable health devices on mental health outcomes: a collaborative study between psychologists, social workers, and nursing staff. J. Int. Cris. Risk Commun. Res. (JICRCR) 7 (2024)
11. Teixeira, E., Fonseca, H., Diniz-Sousa, F., Veras, L., Boppre, G., Oliveira, J., et al.: Wearable devices for physical activity and healthcare monitoring in elderly people: a critical review. Geriatrics **6**(2), 38 (2021)
12. Sharma, A., Badea, M., Tiwari, S., Marty, J.L.: Wearable biosensors: an alternative and practical approach in healthcare and disease monitoring. Molecules **26**(3), 748 (2021)
13. Sandic Spaho, R., Uhrenfeldt, L., Fotis, T., Kymre, I.G.: Wearable devices in palliative care for people 65 years and older: a scoping review. Digit. Health **9**, 20552076231181212 (2023)
14. Prieto-Avalos, G., Cruz-Ramos, N.A., Alor-Hernandez, G., Sánchez-Cervantes, J.L., Rodriguez-Mazahua, L., Guarneros-Nolasco, L.R.: Wearable devices for physical monitoring of heart: a review. Biosensors **12**(5), 292 (2022)
15. Pavithra, L.S., Khurdi, S.: Impact of remote patient monitoring systems on nursing time, healthcare providers, and patient satisfaction in general wards. Cureus **16**(6) (2024)
16. Serrano, L.P., Maita, K.C., Avila, F.R., Torres-Guzman, R.A., Garcia, J.P., Eldaly, A.S., et al.: Benefits and challenges of remote patient monitoring as perceived by health care practitioners: a systematic review. Perm. J. **27**(4), 100 (2023)
17. Tursunbayeva, A., Renkema, M.: Artificial intelligence in health-care: implications for the job design of healthcare professionals. Asia Pac. J. Hum. Resour. **61**(4), 845–887 (2023)
18. Eappen, P., Parker Davidson, K., MacLeod, M., Cousins, A., Gunn, V.: Global healthcare informatics—structural considerations and workforce training challenges and solutions. In: Eappen, P., Vajjhala, N.R. (eds.) Healthcare informatics innovation post Covid-19 pandemic, 1st edn. Routledge, Taylor and Francis & CRC Press (2025)
19. Gunn, V., Villeneuve, M., O'Campo, P., Muntaner, C.: Underfunding of nursing education and the precarious employment conditions of nurses: an exploration of contributing factors, Covid-19 pandemic implications, and structural solutions. In: Conley, H., Koskinen Sandberg, P. (eds.) Handbook on Gender and Public Sector Employment, pp. 321–35. Edward Elgar (2023)

20. Sakphrom, S., Limpiti, T., Funsian, K., Chandhaket, S., Haiges, R., Thinsurat, K.: Intelligent medical system with low-cost wearable monitoring devices to measure basic vital signals of admitted patients. Micromachines **12**(8), 918 (2021)
21. Mirmomeni, M., Fazio, T., von Cavallar, S., Harrer, S.: From wearables to THINKables: artificial intelligence-enabled sensors for health monitoring. In: Wearable Sensors, pp. 339–356. Elsevier (2021)
22. Hariharan, U., Rajkumar, K., Akilan, T., Jeyavel, J.: Smart wearable devices for remote patient monitoring in healthcare 4.0. In: Internet of Medical Things: Remote Healthcare Systems and Applications, pp. 117–135 (2021)
23. Al-Khafajiy, M., Baker, T., Chalmers, C., Asim, M., Kolivand, H., Fahim, M., et al.: Remote health monitoring of elderly through wearable sensors. Multimedia Tools Appl. **78**(17), 24681–24706 (2019). https://doi.org/10.1007/s11042-018-7134-7
24. Sreenilayam, S.P., Ahad, I.U., Nicolosi, V., Brabazon, D.: MXene materials based printed flexible devices for healthcare, biomedical and energy storage applications. Mater. Today **43**, 99–131 (2021)
25. Rajendran, S., Chaudhari, S., Giridhar, S.: Advancements in healthcare using wearable technology. In: Manocha, A.K., Jain, S., Singh, M., Paul, S. (eds.) Computational intelligence in healthcare, pp. 83–104. Springer, Cham (2021)
26. Islam, M.M., Mahmud, S., Muhammad, L.J., Islam, M.R., Nooruddin, S., Ayon, S.I.: Wearable technology to assist the patients infected with novel coronavirus (COVID-19). SN Comput. Sci. **1**, 1–9 (2020)
27. Eappen, P., Olujinmi, T.D.: Telemedicine and digital public health in pandemic times. In: Vajjhala, N.R., Eappen, P. (eds.) Health Informatics and Patient Safety in Times of Crisis, pp. 118–137. IGI Global (2023)
28. Charkviani, M., Barreto, E.F., Pearson, K.K., Amberg, B.M., Amundson, R.H., Bell, S.J., et al.: Development and implementation of an acute kidney injury remote patient monitoring program. Can. J. Kidney Health Dis. **10**, 20543581231192744 (2023)
29. Asensio-Cuesta, S., Sánchez-García, Á., Conejero, J.A., Saez, C., Rivero-Rodriguez, A., García-Gómez, J.M.: Smartphone sensors for monitoring cancer-related quality of life: app design, EORTC QLQ-C30 mapping and feasibility study in healthy subjects. Int. J. Environ. Res. Public Health **16**(3), 461 (2019)
30. Kang, H.S., Exworthy, M.: Wearing the future—wearables to empower users to take greater responsibility for their health and care: scoping review. JMIR Mhealth Uhealth **10**(7), e35684 (2022)
31. Glasby, J., Litchfield, I., Parkinson, S., Hocking, L., Tanner, D., Roe, B., et al.: New and emerging technology for adult social care–the example of home sensors with artificial intelligence (AI) technology. Health Soc. Care Deliv. Res. 1–64 (2023)
32. Jo, A., Coronel, B.D., Coakes, C.E., Mainous, A.G., III.: Is there a benefit to patients using wearable devices such as Fitbit or health apps on mobiles? A systematic review. Am. J. Med. **132**(12), 1394–1400 (2019)
33. Brodeur, A., Gray, D.M., Islam, A., Bhuiyan, S.: A literature review of the economics of Covid-19. SSRN Journal. (2020). https://doi.org/10.2139/ssrn.3636640
34. Keller, S.C., Salinas, A.B., Oladapo-Shittu, O., Cosgrove, S.E., Lewis-Cherry, R., Osei, P., et al.: The case for wearable proximity devices to inform physical distancing among healthcare workers. JAMIA Open **4**(4), ooab095 (2021)
35. Shafik, W.: Smart biomedical devices for smart healthcare. In: Machine Learning Models and Architectures for Biomedical Signal Processing, pp. 421–448. Elsevier (2025)
36. Tavakoli, M., Carriere, J., Torabi, A.: Robotics, smart wearable technologies, and autonomous intelligent systems for healthcare during the COVID-19 pandemic: an analysis of the state of the art and future vision. Adv Intell. Syst. **2**(7), 2000071 (2020)
37. Bootsman, R., Markopoulos, P., Qi, Q., Wang, Q., Timmermans, A.A.A.: Wearable technology for posture monitoring at the workplace. Int. J. Hum. Comput. Stud. **132**, 99–111 (2019)
38. Guk, K., Han, G., Lim, J., Jeong, K., Kang, T., Lim, E.-K., et al.: Evolution of wearable devices with real-time disease monitoring for personalized healthcare. Nanomaterials **9**(6), 813 (2019)

39. Barac, M., Scaletty, S., Hassett, L.C., Stillwell, A., Croarkin, P.E., Chauhan, M., et al.: Wearable technologies for detecting burnout and well-being in health care professionals: scoping review. J. Med. Internet Res. **26**, e50253 (2024)
40. Beniczky, S., Wiebe, S., Jeppesen, J., Tatum, W.O., Brazdil, M., Wang, Y., et al.: Automated seizure detection using wearable devices: a clinical practice guideline of the international league against epilepsy and the international federation of clinical neurophysiology. Clin. Neurophysiol. **132**(5), 1173–1184 (2021)
41. Chaurasia, S.K., Reddy, S.R.N.: State-of-the-art survey on activity recognition and classification using smartphones and wearable sensors. Multimedia Tools Appl. 1–32 (2022)
42. Mao, P., Li, H., Yu, Z.: A review of skin-wearable sensors for non-invasive health monitoring applications. Sensors **23**(7), 3673 (2023)
43. Albahri, O.S., Albahri, A.S., Mohammed, K.I., Zaidan, A.A., Zaidan, B.B., Hashim, M., et al.: Systematic review of real-time remote health monitoring system in triage and priority-based sensor technology: taxonomy, open challenges, motivation and recommendations. J. Med. Syst. **42**, 1–27 (2018)
44. Gluck, S., Lee-anne, S.C., Chapman, M.J., Iwashyna, T.J., Deane, A.M.: A scoping review of use of wearable devices to evaluate outcomes in survivors of critical illness. Crit. Care Resusc. **19**(3), 197–204 (2017)
45. Kekade, S., Hseieh, C.-H., Islam, M.M., Atique, S., Khalfan, A.M., Li, Y.-C., et al.: The usefulness and actual use of wearable devices among the elderly population. Comput. Methods Programs Biomed. **153**, 137–159 (2018)
46. Thomas, E.E., Taylor, M.L., Banbury, A., Snoswell, C.L., Haydon, H.M., Rejas, V.M.G., et al.: Factors influencing the effectiveness of remote patient monitoring interventions: a realist review. BMJ Open **11**(8), e051844 (2021)
47. Qureshi, F.M., Golan, R., Ghomeshi, A., Ramasamy, R.: An update on the use of wearable devices in men's health. World J. Men's Health. **41**(4), 785 (2023)
48. Gustavsson, K., van Diepen, C., Fors, A., Axelsson, M., Bertilsson, M., Hensing, G.: Healthcare professionals' experiences of job satisfaction when providing person-centred care: a systematic review of qualitative studies. BMJ Open **13**(6), e071178 (2023)
49. La Torre, G., Esposito, A., Sciarra, I., Chiappetta, M.: Definition, symptoms and risk of technostress: a systematic review. Int. Arch. Occup. Environ. Health **92**, 13–35 (2019)
50. Califf, C.B., Sarker, S., Sarker, S.: The bright and dark sides of technostress: a mixed-methods study involving healthcare IT. MIS Q **44**(2), 809–856 (2020)
51. Alshammari, M.H., Alenezi, A.: Nursing workforce competencies and job satisfaction: the role of technology integration, self-efficacy, social support, and prior experience. BMC Nurs. **22**(1), 308 (2023)
52. Schall, M.C., Jr., Sesek, R.F., Cavuoto, L.A.: Barriers to the adoption of wearable sensors in the workplace: a survey of occupational safety and health professionals. Hum. Factors **60**(3), 351–362 (2018). https://doi.org/10.1177/0018720817753907
53. Marassi, S., Foldes, M.E.: Deployment of in-body wearable devices in healthcare and at work: closing the EU regulatory gap. Eur. J. Public Health **34**(Supplement_3), ckae144-1109 (2024)
54. Cordeiro, J.V.: Digital technologies and data science as health enablers: an outline of appealing promises and compelling ethical, legal, and social challenges. Front. Med. **8**, 647897 (2021)
55. Krzywdzinski, M., Evers, M., Gerber, C.: Control and flexibility: the use of wearable devices in capital-and labor-intensive work processes. ILR Rev. 00197939241258206 (2024)
56. Ma, M., Li, Y., Gao, L., Xie, Y., Zhang, Y., Wang, Y., et al.: The need for digital health education among next-generation health workers in China: a cross-sectional survey on digital health education. BMC Med. Educ. **23**(1), 541 (2023)
57. Khakurel, J., Melkas, H., Porras, J.: Tapping into the wearable device revolution in the work environment: a systematic review. Inf. Technol. People **31**(3), 791–818 (2018). https://doi.org/10.1108/ITP-03-2017-0076
58. Eappen, P., Gunn, V., Stedman, I.: Transforming healthcare: the role of health informatics and provider perspectives. In: Vajjhala, N.R., Martiri, E., Dalipi, F., Yang, B. (eds.) Artificial Intelligence in Healthcare Information Systems—Security and Privacy Challenges, 1st edn. Springer Nature (2025)

59. Ahammed, M.F., Labu, M.R.: Privacy-preserving data sharing in healthcare: advances in secure multiparty computation. J. Med. Health Stud. **5**(2), 37–47 (2024)
60. Pandit, H.J., Lindquist, J., Krog, G.P. (eds.) Implementing ISO/IEC TS 27560: 2023 Consent Records and Receipts for GDPR and DGA2024: Springer (2024)
61. Obaid, O.I., Salman, S.A.-B.: Security and privacy in IoT-based healthcare systems: a review. Mesop. J. Comput. Sci. **2022**, 29–39 (2022)
62. Goodyear, V.A.: Social media, apps and wearable technologies: navigating ethical dilemmas and procedures. Qual. Res. Sport. Exerc. Health **9**(3), 285–302 (2017)
63. Vajjhala, N.R., Eappen, P. (eds.): An Exploratory Review of Machine Learning and Deep Learning Applications in Healthcare Management. International Conference on Machine Learning and Big Data Analytics 2023. Springer (2023)
64. Mikkonen, K., Ferdinando, H., Sobocinski, M., Kuivila, H., Pramila-Savukoski, S., Vhitehead, T., et al. (eds.): How Does Human-Centred Extended Reality Support Healthcare Students' Learning in Clinical Conditions? 2024. Springer (2024)
65. Fridolfsson, J., Arvidsson, D., Doerks, F., Kreidler, T.J., Grau, S.: Workplace activity classification from shoe-based movement sensors. BMC Biomed. Eng.. **2**, 1–8 (2020)
66. Créquit, P., Mansouri, G., Benchoufi, M., Vivot, A., Ravaud, P.: Mapping of crowdsourcing in health: systematic review. J. Med. Internet Res. **20**(5), e187 (2018)
67. Volpato, L., del Río, C.M., Senn, N., Santiago, D.M.: General practitioners' perceptions of the use of wearable electronic health monitoring devices: qualitative analysis of risks and benefits. JMIR Mhealth Uhealth **9**(8), e23896 (2021)
68. Vivarelli, C., Censi, F., Calcagnini, G., Falsaperla, R., Mattei, E.: Risk assessment for workers with wearable medical devices exposed to electromagnetic fields. Health Phys. **127**(2), 269–275 (2024). https://doi.org/10.1097/HP.0000000000001798
69. Gonçalves, J., Leitão, L., Carvalho, V. (eds.): Analysis of the Reliability and Accuracy of a Wearable Device: Comparative Study with a Certified Clinical Device. IEEE (2017)
70. Dinh-Le, C., Chuang, R., Chokshi, S., Mann, D.: Wearable health technology and electronic health record integration: scoping review and future directions. JMIR Mhealth Uhealth **7**(9), e12861 (2019)
71. Tanaka, M., Ishii, S., Matsuoka, A., Tanabe, S., Matsunaga, S., Rahmani, A., et al.: Perspectives of Japanese elders and their healthcare providers on use of wearable technology to monitor their health at home: a qualitative exploration. Int. J. Nurs. Stud. **152**, 104691 (2024)
72. Løvseth, L.T., de Lange, A.H.: Integrating the Organization of Health Services, Worker Wellbeing and Quality of care: Towards Healthy Healthcare. Springer (2020)
73. Løvseth, L.T., de Lange, A.H.: Integrating organisation of healthcare services, workers' well-being, and quality of care: an introduction to the system-based perspective of healthy healthcare. In: Integrating the Organization of Health Services, Worker Wellbeing and Quality of Care: Towards Healthy Healthcare, pp. 3–14. Springer (2020)
74. Van der Zwart, J., Pilosof, N.P.: Evidence-based design for healthcare work environments. In: Integrating the Organization of Health Services, Worker Wellbeing and Quality of Care: Towards Healthy Healthcare, pp. 245–262 (2020)
75. Cheung, M.L., Chau, K.Y., Lam, M.H.S., Tse, G., Ho, K.Y., Flint, S.W., et al.: Examining consumers' adoption of wearable healthcare technology: the role of health attributes. Int. J. Environ. Res. Public Health **16**(13), 2257 (2019)
76. Smuck, M., Odonkor, C.A., Wilt, J.K., Schmidt, N., Swiernik, M.A.: The emerging clinical role of wearables: factors for successful implementation in healthcare. NPJ Digit. Med. **4**(1), 1–8 (2021)
77. Hilty, D.M., Armstrong, C.M., Edwards-Stewart, A., Gentry, M.T., Luxton, D.D., Krupinski, E.A.: Sensor, wearable, and remote patient monitoring competencies for clinical care and training: scoping review. J. Technol. Behav. Sci.. **6**(2), 252–277 (2021)
78. Ozanne, A., Johansson, D., Hällgren Graneheim, U., Malmgren, K., Bergquist, F., Alt, M.M.: Wearables in epilepsy and Parkinson's disease—a focus group study. Acta Neurol. Scand. **137**(2), 188–194 (2018)

79. Tursunbayeva, A., Bunduchi, R., Franco, M., Pagliari, C.: Human resource information systems in health care: a systematic evidence review. J. Am. Med. Inform. Assoc. (JAMIA) **24**(3), 633–654 (2017). https://doi.org/10.1093/jamia/ocw141
80. Idoga, E.P., Adamu, A.I.: Understanding smart wearable sensors technology: impact on human health and fitness. J. Appl. Sci. Environ. Manage. **24**(7), 1261–1265 (2020)
81. Huang, J.-D., Wang, J., Ramsey, E., Leavey, G., Chico, T.J.A., Condell, J.: Applying artificial intelligence to wearable sensor data to diagnose and predict cardiovascular disease: a review. Sensors **22**(20), 8002 (2022)
82. Sabry, F., Eltaras, T., Labda, W., Alzoubi, K., Malluhi, Q.: Machine learning for healthcare wearable devices: the big picture. J. Healthc. Eng. **2022**(1), 4653923 (2022)
83. Bokharaei Nia, M., Afshar Kazemi, M., Valmohammadi, C., Abbaspour, G.: Wearable IoT intelligent recommender framework for a smarter healthcare approach. Libr. Hi Tech **41**(4), 1238–1261 (2023)
84. Turner, K., Jo, A., Wei, G., Tabriz, A.A., Clary, A., Jim, H.S.L.: Sharing patient-generated data with healthcare providers: findings from a 2019 national survey. J. Am. Med. Inform. Assoc. **28**(2), 371–376 (2021)
85. Azodo, I., Williams, R., Sheikh, A., Cresswell, K.: Opportunities and challenges surrounding the use of data from wearable sensor devices in health care: qualitative interview study. J. Med. Internet Res. **22**(10), e19542 (2020)
86. Ferguson, C., Hickman, L.D., Turkmani, S., Breen, P., Gargiulo, G., Inglis, S.C.: "Wearables only work on patients that wear them": barriers and facilitators to the adoption of wearable cardiac monitoring technologies. Cardiovasc. Digit. Health J. **2**(2), 137–147 (2021)
87. White, M., Pizzetta, C., Davidson, E., Hines, A., Azevedo, M., Ikem, F., et al.: Mississippi church leaders' perceptions of challenges and barriers to the use of consumer wearables among community members. AIMS Public Health **10**(4), 775 (2023)
88. Johansson, D., Malmgren, K., Alt, M.M.: Wearable sensors for clinical applications in epilepsy, Parkinson's disease, and stroke: a mixed-methods systematic review. J. Neurol. **265**, 1740–1752 (2018)

Virginia Gunn is an Associate Professor at Cape Breton University and affiliated researcher at Karolinska Institutet. She specializes in public health, social determinants of health, and the impacts of AI and work digitization on healthcare systems. Dr. Gunn collaborates internationally with teams from Australia, Europe, and North America. Her most recent work investigates race, migration, and employment as health equity determinants.

Philip Eappen is an Assistant Professor at Cape Breton University. He has more than 15 years of experience in healthcare leadership and teaching. Dr. Eappen holds a DHA from Central Michigan University and has published extensively in healthcare innovation. He has served as Director of Clinical Services at the Breton Ability Center and held senior positions at universities in Canada and Nigeria. He serves on several healthcare boards, including AAMAC and NRBDO.

Hikmat Singh Brar is a foreign-trained physician and Research Assistant at Cape Breton University. He earned his MBBS from Dayanand Medical College and completed residency training in cardiology. His capstone project addresses Canada's GP shortage and proposes structural reforms. His interests include health systems planning and sustainable primary care.

Emma Brulin is an Associate Professor at the Institute of Environmental Medicine, Karolinska Institutet. She researches psychosocial work conditions in healthcare and implementation of sustainable practices. Her publications span moral distress, job stress, and organizational change in healthcare systems.

Carles Muntaner is a Professor at the University of Toronto and Universitat Pompeu Fabra. A social epidemiologist, he studies the intersections of class, race, colonialism, and health inequities. He has received multiple honors including the Wade Hampton Frost Award and is a fellow of the Canadian Academy of Health Sciences. His work critically evaluates employment systems and their impact on mental and physical health.

Patient and Participant Engagement and Involvement in the Co-Design of Digital Health Interventions

Pauline Frank and William R. Young

Abstract The integration of patient and public involvement (PPI) in the development of digital health interventions (DHI) is essential for ensuring usability, engagement, and long-term adoption. While human-centered design (HCD) has been widely applied, its inconsistent implementation has led to interventions that fail to meet real-world patient needs, contributing to low adherence and poor scalability. This chapter advocates for a shift towards human-centered participatory design (HCPD) as a pragmatic method for PPI in DHI development, emphasizing early and sustained stakeholder engagement, shared decision-making, and contextual integration. We outline key principles for effective PPI, including defining problems from patient perspectives, mapping and involving diverse stakeholders, and aligning design with implementation frameworks. A structured, stepwise approach is presented for co-designing, prototyping, deploying, and scaling DHIs while addressing common barriers such as tokenism, misalignment of stakeholder priorities, and digital literacy challenges. By embedding PPI at every stage, DHIs can be designed to enhance patient empowerment, adherence, and clinical utility. A case study of a long-running Parkinson's disease initiative is provided to demonstrate the need for meaningful PPI in real-world settings. This chapter thus provides a practical roadmap for researchers, developers, and healthcare professionals to integrate PPI effectively, ensuring that digital health innovations are not only technologically sound but also responsive to the needs and preferences of the populations they serve.

Keywords Patient engagement · Participatory involvement · HealthTech co-design

P. Frank (✉)
Patient Co-Next, Zürich, Switzerland
e-mail: pauline@patient-co-next.com

W. R. Young
University of Exeter, Exeter, UK

P. Eappen et al. (eds.), *Remote Monitoring and Wearable Devices in Healthcare*,
Information Systems Engineering and Management 63,
https://doi.org/10.1007/978-3-031-98897-4_7

1 Introduction

The concept of involving end-users in the design of technology is not new, evolving over the past 50 years from simply asking users what they want to having them as an equal partner in the design process. However, the practice of human-centered design (HCD) is not applied consistently or with clear methodology in practice. This can result in products being developed in a manner that does not address the actual end-user needs, resulting in poor market uptake, low engagement or attrition. It can also result in significant waste of resources and propagate an endless state of 'pilot-itis' in healthcare technology—innovation which begins and ends with a pilot or beta model without progressing to more advanced stages of development [1].

This chapter aims to provide key principles to improve patient and participant involvement (PPI) in healthcare technology based on learnings from published research and practical experience, to prepare teams to embrace PPI in the co-design of future digital health interventions (DHI) for improved success in research, development and implementation.

2 Defining Patient Engagement, Co-design and the Role of PPI

In past decades, the term 'patient engagement' has carried multiple meanings including reference to activating patients in the use of digital tools used to manage a given health condition [2, 3]. However, in general healthcare research and drug development, the terms 'patient engagement' and 'co-design' are often used interchangeably to encompass the notion that patients should not only be considered as end-users of a product, but also as equal partners in every stage of the research and design process [4].

The general concept of involving public and patient stakeholders in the research process and DHI is not new. In the 1970s there was recognition of the importance of the patient voice in health innovations. However, it was not until the 1990s that the term PPI became widely accepted as an approach to advocating greater accountability, transparency and responsiveness to the needs and preferences of end-users. The understanding and use of the term 'co-design' also appears to be inconsistent. The term is used to represent a broad range of levels at which end-users are invited and supported to participate, leading to a variability in what involvement actually looks like in the design process, the measures used and outcomes assessed by project [5].

To many people, especially those involved in the technological side of DHI, PPI might be considered as an unnecessary and inefficient process designed to satisfy funders and regulators. It is also a common reality that PPI can be carried out in a tokenistic fashion; a tick-box exercise to meet external project requirements.

The aim of this chapter is to advocate for a pragmatic approach to PPI, where substantial benefits can be achieved by focusing on core principles. We also aim to demonstrate that PPI should not be considered as a cumbersome or non-essential process, but rather an integral part of any DHI development to enable the needs of stakeholders to be established and prioritized. On a practical level, this means that stakeholders are given an equal opportunity to engage in a collaborative process to share their experiences, values (personal and cultural), expectations and insights to inform or evaluate new innovations.

This chapter will provide:

- Key principles to organize PPI activities in a way that maximizes efficiency and beneficial impacts for the DHI.
- Key considerations to conduct and adapt PPI based on the applied context.
- A case study of a real and ongoing DHI initiative in Parkinson's disease with PPI.

3 Defining the Problem to Solve with PPI

Despite good intentions, technology development teams often miss certain steps to truly engage with end-users. Although they are adept at defining the product vision, they only know what they know—which may not include knowledge of real user perspectives. There is often a tendency to be led by this creative vision and therefore view additional input or suggestions for alteration as potential criticism or unwanted disruption. The goal is not just to produce a digital tool but, rather, to solve or improve a healthcare challenge important to patients, their families, healthcare providers (HCPs), health system administrators, and to society. As such, any potential shortcomings in understanding end-user perspectives could result in critical problems that may not be identified until such perspectives are later voiced further downstream in the development process.

3.1 Key Errors in DHI Development Relating to Lack of End-User Engagement

Aim to solve a problem which has not been derived from the actual needs of target end-users. The problem and idea to solve it has come from the developers' own perceptions or experiences, or from a proxy for the end-user (e.g., only physicians are asked what they think would be the requirements for a wearable to be worn by patients). There is a lack of attention to the patient or participant's 'lived experience' of a disease or health condition, and lack of tailoring to their daily lives, when designing the DHI [6].

Limited involvement of end-users in the full design process. Patients may be asked in initial research steps via surveys, interviews, or focus groups about their disease lived experience, expectations and preferences for a DHI. They may be asked to test the prototype once developed but are not involved in co-design steps [7].

Focus only on a laboratory testing environment, and not actively consider real-world implementation. Surrogate measures are used rather than end-user defined measures, potentially because they are simpler to measure in a laboratory environment and easier to show that the DHI is 'successful' (e.g., number of minutes spent per day on a mobile application vs user satisfaction score). Laboratory-based methods may have increased sensitivity, making it easier to show statistically significant differences. It is also easier to demonstrate differences when testing is not conducted within the complexities and distractions encountered in daily life.

3.2 Implications of Inappropriate End-User Engagement

The result of inappropriate end-user engagement includes:

- Lack of uptake and adoption.
- High attrition rate (e.g., wearables discontinued or abandoned after 6 months by 1 in 3 users, after 2 weeks in those 70 years or older).
- Lack of adherence which means poor data quality and lower impact for use in diagnosis and treatment decisions.
- Lack of scale up and long-term engagement.
- Expected benefits of DHI are not met.

It is therefore critical as part of the development process, to engage adequately with all relevant stakeholders to define and agree on the problem(s) to be solved as a first milestone.

4 Downstream, Midstream and Upstream Stakeholders

It is a good exercise to categorize stakeholders into groups based on their roles in the innovation process, and their level of influence. Each group has unique contributions at various stages of development, deployment and usage of the DHI. One method is to group stakeholder streams and determine the priorities of each category in terms of what they want from a DHI.

Table 1 Potential downstream stakeholders, their roles and needs in DHI innovation

Role	What they want	Examples
Research, design, utilization, feedback, impact realization	*Example for Patients* • Prevention of or minimize worsening of disease • Preserve QoL via reduced burden to travel to clinics • DHI that helps reduce but not replace in-person interaction with HCPs • Preferences, values and data privacy considered • Better self-care and symptom management • Access to information • Improve communication with HCPs • Easy to use, discrete and comfortable tech/wearables • Impairments accommodated (vision, hearing, etc.)	Patients, caregivers, patient navigator, community health worker

4.1 Downstream Stakeholders

Downstream stakeholders are the end-users and those directly impacted by DHI deployment and outcomes. Patients are typically the primary users of monitoring devices and wearables, e.g., a diabetic patient uses a continuous glucose monitor (CGM) connected to a mobile application. They can also include caregivers and families who support patients in using the digital tools and interpreting data, e.g., an adult child uses remote monitoring alerts to support compliance to medication intake by their elderly parent (Table 1).

4.2 Midstream Stakeholders

Midstream stakeholders are primarily responsible for the success of the DHI in terms of design, integration into clinical workflows and systems, deployment, adoption and scale up. HCPs are important midstream stakeholders to ensure that the DHI and data generated are useful and works within existing clinical processes and infrastructure, thereby promoting connectivity between patient and health system. They may also be downstream stakeholders as users of patient data generated from DHIs to aid disease and treatment management (Table 2).

Table 2 Potential midstream stakeholders, their roles and needs in DHI innovation

Role	What they want	Examples
Research, funding, ethics, insights, design, deployment, integration, scaling	*Example for Physicians* • Preferences and ways of working integrated into the design • Streamlined care and communication with patients • Improve patient medication adherence • Ability to better monitor symptoms, including mental health, of patient • Time and cost savings • Reliable clinical data • Not an additional workload or data burden • Works within the limitations of their IT infrastructure • Clear guidelines to manage potential legal liability (e.g., acting on patient data from DHI) • Customizable—more prescriptive for less experienced clinicians, less prescriptive within bounds of evidence for more experienced clinicians	HCPs, researchers, developers, IT experts, system administrators

4.3 Upstream Stakeholders

Upstream stakeholders are important for not only getting the DHI project started, but also when planning for scale up, broader adoption, and sustainability. They can support funding models which go beyond pilot stage, e.g., reimbursing use of the DHI in clinical practice. They essentially provide the health system enablers for implementation, and their support and buy-in is thereby essential for a long-term outlook (Table 3).

4.4 Value of Stakeholder Mapping

The stakeholder mapping shown in the tables above provides a structure for better understanding of all the key players in DHI development and deployment, some common barriers which they face or perceive (as individuals or systemic) and their potential motivations for adoption or change. The needs of different stakeholders may

Table 3 Potential upstream stakeholders, their roles and needs in DHI innovation

Role	What they want	Examples
Funding, regulation, feedback	• Improve use of and adherence to guideline directed medication therapy • Reduce costs and implications of poor patient outcomes, e.g., hospitalization • DHI operating within legal, regulatory, and ethical guidelines	Funders, payers, policy makers

align or conflict. For example, HCPs may be focused on physical symptoms whereas patients place more emphasis on emotional/spiritual (holistic) care needs. Where conflict emerges, a well-structured dialogue is required to inform prioritization within DHI [8].

Additionally, in certain diseases like Parkinson's, DHI should not only be for individual patient use but also aim for a model of care that is collaborative and integrated with involvement of caregivers/family and HCPs [9]. The case study at the end of the chapter provides an example of a real-world case scenario.

Thus, DHI designed with all stakeholders mapped out, involved, and their needs addressed can ensure the DHI developed is effective, widely adopted, and impactful for patients and the broader health system. It can better help patients overcome barriers to access support and services within the health systems which they live—many of which may be disconnected or fragmented, whereby technology offers immense opportunities for improved coordinated care.

5 Designing for Implementation

In the field of implementation research, it is acknowledged that the average time to implement evidence into practice is 18 years. If the appropriate implementation strategies and frameworks are applied, the timeframe and likelihood of success to implement a health intervention into practice are improved. Within implementation research, the environmental context (human, socioeconomic, cultural and technological factors) is important to understand, as it can influence the outcomes of the DHI. Technology cannot be successfully implemented if the context, workflows and system integration are not considered. Attention should also be paid to whether low-technology or technology-free solutions may be more appropriate or complementary to a more complex DHI solution [1].

PPI can help address this evidence to practice paradigm when developing DHI, enhancing the real-world application and use of DHI. What can the key benefits look like from an implementation research perspective?

5.1 Potential Benefits of PPI in the Implementation of DHI

Practical Feasibility. The DHI works within the existing infrastructure and workflows where it is intended to be used.

Consider a digital health application that requires good continuous internet connectivity—implementation will not succeed in remote or underserved areas, and a good implementation assessment will be able to identify areas more appropriate for implementation or the need to design a tool that works also off-line.

Adoption by End-Users. Patients, HCPs and other stakeholders are more willing to adopt DHI that is easy to use, affordable, addresses their real pain points, and aligns with existing systems and workflows. PPI thus helps identify what developers may not already know and which, once the unforeseen issues are uncovered, may actually seem quite obvious yet not considered prior to stakeholder insights. This issue highlights the importance of researchers and developers being aware that there are considerable limitations to their understanding unless they are informed by end-user insights.

Consider a telehealth platform to enable patient monitoring remotely—when integrated seamlessly with electronic health records (EHRs), lab results and physician notes, the administrative burden is reduced and clinicians will be motivated to adopt it in clinical practice. Any additional time required for use of the DHI or its data may require clinician renumeration.

Lower Cost and Waste. Although requiring more effort initially, approaching co-design with implementation in mind minimizes the risk of redundant features, missed issues or the need for significant changes. Thus, ultimately saving time and money.

Consider a wearable device for older patients requiring 24/7 cardiac monitoring—robust steps in PPI would ensure the wearable is discrete, easy to use, comfortable and the benefits are tangible to motivate them to use it. The wrong design, poor instructions and lack of user support would mean poor patient adherence and efforts to retrain and/or redesign [10].

Desired Behavior Change. DHI implementation requires a strategy to enable targeted behavior change in users. PPI can be used to select the right enablers (tool features, education, etc.) to support adoption.

Consider the deployment of a disease management mobile application that incorporates behavioral strategies such as daily symptom logging, goal setting, personalized messages with educational and motivational content, prompting or reminders, gamification and incentivization [10].

Aligned with Regulations and Ethics. Patient concerns about data privacy and consent are appropriately addressed, including compliance to laws like General Data Protection Regulation (GDPR) or Health Insurance Portability and Accountability Act (HIPAA), ensuring approval processes are smooth when entering the market and the DHI is trusted among stakeholders.

Consider a health management application which has data privacy measures built-in to its design and workflow from the start—upon deployment there will be less issues and roadblocks.

Impactful Outcomes Measured. Measures of success agreed with stakeholders are built into the DHI design, making it easier to demonstrate value after deployment.

Consider a mobile application to manage diabetes which has integrated analytics in a format co-designed with HCPs and payers—the data can be easily accessed by these stakeholders to assess and demonstrate return on investment, patient outcomes measures, etc.

Scalability and Sustainability. Resources, costs and infrastructure requirements for DHI implementation are evaluated and validated with stakeholders, that can be extrapolated to other settings to scale up use over the long-term.

Consider a patient-physician application developed with good PPI practices for implementation—a tested and accepted onboarding approach can be deployed faster to other similar healthcare settings/users, with adaptation for different settings requiring less overall effort, and stakeholder buy-in ensuring sustained support and investment.

5.2 Aligning Terminologies

Rather than invent new terminologies, we will align with existing and evolving terminologies used in the context of creative design methodologies to illustrate the requirements for PPI in DHI development. When we use the term 'patient and participant involvement' we mean the same approach as what is known as 'human-centered participatory design' which is an evolution of what we have previously discussed 'human-centered design'.

HCPD enables the design of DHI to follow the principles of implementation research.

6 Moving from Human-Centered Design to Human-Centered Participatory Design

Human-centered design (HCD) is a creative design methodology developed in the 1970s to facilitate innovative solution development processes. It has been widely used in engineering and social sciences, and only more recently in healthcare intervention development. There are > 100 HCD methodologies described, all with the underpinning purpose of understanding end-user needs through their engagement in the design process. If conducted well in healthcare innovation, it fosters the concept of person-centered care by considering their needs, values and preferences in a holistic manner [11].

However, as highlighted, the application of HCD has been inconsistent and less than ideal in many cases. To shift the paradigm towards a better design process of DHI, we have discussed and reiterated these principles:

- Defining the problem or challenge from the perspective of end-users.
- End-users (patients, caregivers, families and HCPs) as true partners in co-design.
- Complete mapping of stakeholders, their roles and needs understood.

These are the foundations to move PPI towards the model of human-centered *participatory* design (HCPD) which is a combination of HCD and participatory design (PD), where participation goes beyond individuals (patients, caregivers, families and HCPs) and includes those who are part of the community or health system where the patient receives care. This enables assessment of the local context in which a DHI may be a key part of the solution to a defined problem or challenge, and requires good understanding and integration into the local context to be successful.

Although HCD and HCPD are both user-focused approaches which aim to understand and address their needs, preferences and experiences, they differ in the degree of stakeholder involvement and methods used throughout the design process. Below is a summary of the key differences (Table 4).

The intent of HCD is to design with empathy to develop solutions which are user-friendly and effective. The developer is the primary decision-maker. HCPD is an evolution of HCD, whereby users and stakeholders are actively involved as co-creators in each stage of the design process—with an emphasis on collaboration and shared decision-making. Users are not just subjects of research but, rather, are treated as equal partners in the design.

Insights from users are gathered in HCD, then solutions are designed for them based on interpretation of the insights by the developers. Besides this, the role of users is limited to testing prototypes or validating solutions. Whereas in HCPD, co-design

Table 4 Comparison of HCD versus HCPD approaches

Aspect	HCD	HCPD
Focus	Designing *for* users	Designing *with* users
User role	Participants in research (mainly for insights) and testing	Same as HCD *plus* co-creators and decision-makers
Level of engagement	Periodic	Iterative
Methods	Observation, interviews, surveys, usability testing	Same as HCD *plus* co-design workshops and participatory prototyping
Control	Developers lead	Shared decision-making between users and developers
Outcom	User-friendly solutions	Collaborative solutions aligned with user values and preferences

happens with users as active stakeholders in ideation, prototyping and decision-making—instilling empowerment, inclusion, and co-ownership of the outcomes in users.

The level of engagement with users in HCD is controlled by developers, and aligned with periods only when their insights or testing is required. In comparison, engagement of users in HCPD is continuous and spans the design process end-to-end, from problem definition to solution implementation. Power is balanced and shared between developers and users.

Although the methods used in HCD and HCPD to get insights from users may be the same (ethnography, interviews, journey mapping, personas, etc.), in HCPD the users are also involved in workshops to co-design, ideate, build prototypes, refine solutions, and assess feedback. The outcomes for both methods are solutions that are user-friendly and address user needs, although additionally, HCPD solutions are more likely to be better aligned with stakeholders' values and system/societal needs, leading to higher adoption and long-term engagement.

Participatory frameworks that may be used include [1]:

- Participatory action research—uses experience-based knowledge of individuals
- Community-based participatory research—to enable community capacity building
- Experience-based co-design—to understand the current healthcare system and services.

7 Perceived Obstacles to Incorporating True Co-Design

It may at this point, seem overwhelming to partner with end-users and other stakeholders in the manner described for HCPD. Questions from development teams already feeling pressure may include:

> Is it really necessary for all DHI development?
>
> How can I find end-users who are willing and able to be involved in what seems to be a lengthy and time-intensive process?
>
> Will I lose control of the scope of my project?
>
> Will I require a lot more time and budget?

There are no simple answers to these commonly asked questions. Our advice is to have a broadly structured process in place for HCPD/PPI with simple core steps and importantly, prepare to learn and be agile.

8 Putting PPI (or HCPD) into Practice

Below are the core steps for conducting HCPD/PPI in a DHI innovation process. The key objectives and outputs are outlined for each step, followed by recommended activities on how to achieve them.

(i) **Exploration and Scoping**

Objectives:

Define the problem, identify team and stakeholders, set the foundation for collaboration.

Outputs:

Clear problem statement.

Defined project scope and aligned objectives.

Initial stakeholder engagement and buy-in.

Recommended Activities:

- Conduct a literature review to identify existing knowledge—learnings, good practices.
- Select the DHI core team, co-design facilitators and methods of engagement.
- Conduct stakeholder mapping to identify all relevant participants—downstream, midstream, upstream.
- Clarify roles and responsibilities and collaboration model—build trust with participants.
- Gather broad perspectives on needs, challenges and priorities from initial discussions and workshops.
- Define the problem and scope (high level, open for discussion), propose initial objectives.

(ii) **Co-research and Understanding**

Objectives:

Deeply understand the needs, experience and context of stakeholders.

Outputs:

Comprehensive insights into stakeholder needs and pain points, including diverse and dissimilar views.

Clarification on terminology or lexicon appropriate for patients (can be different to clinical and HCP accepted language).

User personas and journey maps.

Clear understanding of the contextual environment (cultural, technical, regulatory, etc.) to enable good implementation planning at the community level.

Recommended Activities:

- Conduct qualitative research to gain insights from broad and representative stakeholder groups and sub-groups—semi-structured interviews, focus groups, etc.

- Conduct participatory research to identify stakeholders' care pathways, workflows, value drivers and pain points—observational studies, ethnography, etc.
- Analyze data from research, identify themes and validate the findings with participants.
- Co-create journey maps or user personas to represent key insights.
- Explore existing solutions and identify gaps.

(iii) **Ideation and Co-creation**

Objectives:

Generate ideas collaboratively to address the identified needs and challenges.

Outputs:

A portfolio of ideas and initial concepts.

Stakeholder agreement on prioritized solutions for further development.

Recommended Activities:

- Develop an exploratory prototype for initial design meeting (an icebreaker for the stakeholder discussions), with clearly defined target user group(s), e.g., a DHI to be developed for stage 1–3 Parkinson's patients.
- Facilitate brainstorming sessions, design sprints, or innovation workshops with stakeholders to build on the exploratory prototype.
- Use tools like sketches, storyboards, mock-ups and mapping to visualize concepts as they evolve.
- Use facilitation methods to encourage diverse stakeholders to contribute their ideas and solutions based on their experiences, e.g., ask a series of "how might we" questions or design possible future scenarios [9, 11].
- Prioritize ideas based on feasibility, viability, desirability, impact and alignment with user needs—consider using methods like a 3-point Likert scale and comparing patient versus HCP preferences.

(iv) **Prototyping and Participatory Testing**

Objectives:

Create tangible prototypes and test them with stakeholders to refine the solution.

Outputs:

Refined prototypes with user-driven design enhancements.

Documentation of user feedback and prototype iteration history.

Recommended Activities:

- Develop low-fidelity prototypes (alpha) to gather early feedback—wireframes, click-through models, etc.
- Iterate prototypes based on stakeholder input.
- Conduct user testing sessions where stakeholders interact with the prototypes and provide feedback.
- Incorporate feedback from user testing to create high-fidelity versions (beta).

(v) **Implementation Planning**

Objectives:

Prepare for real-world deployment of the co-designed DHI.

Outputs:

Comprehensive implementation roadmap.

User-centered training materials and resources.

Strategies for resistance, troubleshooting, scaling, and sustaining the solution.

Recommended Activities:

- Co-develop and implementation plan with stakeholders to map out workflows, training, technical integration, etc.
- Identify potential barriers to adoption (e.g., resource constraints, data privacy or regulatory issues) and co-create strategies to overcome them—consider local setting and user sub-groups.
- Select an appropriate local setting for piloting where the need is high (not just a nice to have) and the basic requirements for an implementation program exist and can be added with minimal effort—experts and data can assist.
- Engage stakeholders to plan for deployment, scalability, interoperability, and sustainability.

(vi) **Deployment and Real-World Testing**

Objectives:

Launch the solution in a controlled environment and evaluate its real-world performance.

Outputs:

Insights from real-world settings to further refine the solution.

Evidence of impact, e.g., improved patient health outcomes, time efficiencies (for HCP in workflow, patient in accessing care needs), satisfaction with user experience of the solution.

Recommended Activities

- Pilot the solution in real-world settings.
- Collect feedback through surveys, interviews, or focus groups to assess usability, satisfaction and effectiveness.
- Identify technical or operational challenges.

(vii) **Reflection and Iteration**

Objectives:

Ensure continuous improvement based on real-world performance and evolving needs.

Outputs:

Enhanced versions of the solution.

Documented lessons learnt.

Stakeholder satisfaction and ongoing engagement.

Recommended Activities:

- Evaluate post-implementation feedback and learnings with stakeholders.
- Measure success of the deployment and implementation using KPIs, analytics, and stakeholder feedback (quantitative and qualitative).
- Be ready to update the solution based on continuous feedback from end-users, evolving needs or challenges.

(viii) **Scaling and Sustaining**

Objectives:

Expand the reach and optimize the long-term use of the solution.

Outputs:

Scaled ongoing implementations adapted to new local settings.

Smooth process established for collaboration, feedback, and solution improvement.

Sustainability strategies, e.g., cost sharing, system integration.

Recommended Activities:

- Co-develop plans for scaling up to more users and expanding to new user groups, healthcare settings or geographies.
- Identify opportunities for future funding, partnerships or support models with stakeholders.
- Continue collaboration model.

9 Challenges of PPI and How to Address Them

Transitioning industry practices from developing a product or solution *for* patients to developing *with* patients, is still a relatively new and broad concept. In many cases, these activities are not reported transparently, thus limiting the learnings that can be derived from what stakeholders felt worked well or otherwise, in addition to perceptions around the value of HCPD. This is the first challenge of PPI—the belief among developers that it is necessary. However, there is growing momentum to change this as patients, their families, patient advocates and organizations align as professional bodies and actively express their expectations to be involved as true collaborators and partners, not only as end-users. As more developers are seen to collaborate with patients, others tend to jump on board as well, if only to show others that they are also 'patient-centered'.

Once this initial challenge is overcome, there are other key issues to proactively resolve.

9.1 The Dilemma with Payments

When developers engage with HCPs or other professionals in collaborative projects, appropriate compensation is expected for the time and expertise provided. This has not always been clear with regard to patient contributors, if and how much they should be paid. Although there has been positive change in this area with companies setting 'fair market' rates to compensate patients and caregivers for their time working collaboratively on projects, the situation is still not perfect with cases cited of no offers of compensation or a lack of transparency in the rates paid. We advise having legal and compliance teams look into the local fair market rates for all stakeholders involved in the co-design process, and clarify at the project start, the appropriate compensation per stakeholder type. Contracts should be in place with clear description of expectations, responsibilities, what will be paid and when.

9.2 Measurement of Impact

Measurement of the impact of PPI in DHI has not been standardized, with a lack of consensus on what constitutes 'meaningful' PPI. An overemphasis on quantifying the outcomes is a factor in PPI at risk of remaining a box-ticking exercise, and forfeits the greater benefits of true social participation in design. Different approaches to the extent of PPI used means a wide range in potential benefits when earlier vs later involvement is deployed, i.e., integrated vs tokenistic PPI. Efforts to standardize the approach and timing of PPI will enable an ability to measure impact across different areas.

Until then, good documentation and transparent reporting of how PPI has influenced DHI development is the best means to demonstrate impact [12]. For example:

Clearly describe the PPI approach—type and number of participants, which stages they were involved in, nature of their involvement, etc.

How did it shape the objectives and product/service design?

How did it influence design reiterations?

How did it influence the selection of target end users and locations for piloting?

How did it influence the selection of measures of user uptake/acceptance?

9.3 Managing Diverse Perspectives

Diverse perspectives are a strength of HCPD, however, when they are so diverse that stakeholders have difficulty aligning priorities, it becomes a challenge. Using an approach like a Delphi method may be useful to drive consensus. Alternatively, quantitative methods (e.g., survey) may be used to better clarify the importance or ranking of different options to a larger group of end-users or different sub-groups of end-users who have not been involved in the co-design process.

9.4 Digital Literacy

There is a general lack of digital literacy amongst patients and HCPs, which needs to be addressed to make them more effective co-design participants and end-users of DHI solutions. Specifically, ethnic minorities, older adults and patients of low socioeconomic status have the highest risk of low digital literacy and thus, likely to have greater difficulty accessing healthcare delivered by digital health [13]. This has important implications when considering diversity, inclusion and representation.

Education campaigns on digital literacy may be warranted to on-board participants and users to technology, prior to the development and deployment of the DHI. The understanding of data privacy is an important component of digital literacy to be taught.

10 Benefits for Diversity, Inclusion and Representation

With the increasing focus on ensuring representation when developing healthcare interventions, we believe that following the principles of HCPD will support diversity and inclusion in DHI development.

Ensuring diversity can be done via purposive sampling when selecting participants for research insights and testing with either HCD or HCPD methods, ensuring representation of gender, age, disease stage, ethnicity/race, language, culture, etc. However, HCPD supports better inclusivity as diverse end-users can help guide the actual use of the diverse range of insights. There is a lower risk of developer bias in selecting what they think is relevant according to their priorities, and ignoring what seems more trivial or complex.

PPI also supports the design of DHI solutions which fit specific cultural, social, and environmental contexts of end-users; and considers local practices, norms, and constraints to ensure relevance and acceptance. For example, in certain locations the communities are built on strong connections between the citizens and local organizations such as religious or cultural groups. DHI implementation would be more successful if it had a community level approach versus a top-down approach from government or academic institutions [14].

Inclusive design also accommodates diversity by considering the characteristics and needs of different user groups whose capabilities vary, e.g., those with lower tactile functionality in movement disorders or those with attention disorders who require a different approach to digital engagement [4]. Participants can bring fresh ideas and creative perspectives that designers, who do not have the lived experience of the disease or of obtaining care in that local context, might not consider on their own.

11 Case Study: DHI for Parkinson's Disease and Self-Management of Gait Disorders

11.1 Overview of Applied Context

Parkinson's disease (PD) is a chronic and progressive neurological disorder of the nervous system. Symptoms include tremor, stiffness, slowing of movement, trouble with balance that raises the risk of falls, masked facial expression, and slurred speech. Symptoms worsen over time, and while some medicines and surgical interventions might help to lessen certain symptoms, no cure is currently available [15].

Freezing of gait (FOG) is widely regarded as one of the most debilitating symptoms of PD. It is described as the sensation that one's feet are glued to the floor. FOG is strongly associated with falls and reduced quality of life. Unfortunately, FOG is often resistant to pharmacological intervention, therefore placing an emphasis on physical therapy and DHIs designed to promote self-management of symptoms.

There is a well-established phenomenon whereby people with PD (including those with FOG) can employ sensory tricks and strategies to improve balance and walking. It is thought that the provision of external sensory movement cues help bypass areas of the brain compromised by PD and, rather, rely on relatively unaffected networks. The following case study describes a program of research led by author WY to develop DHI in the context of FOG. The aim of the case study is to highlight common mistakes in addition to the challenges and benefits that come with engagement in PPI.

11.2 Original Project Background

A mobile phone application was developed to play continuous auditory stimuli that had previously been shown in laboratory settings to improve gait characteristics and reduce FOG severity in people with PD [16–18]. The researchers felt assured in assuming the validity of the targeted applied problem. Not only were the negative implications of FOG well-articulated in the research literature, but Parkinson's UK (a charity supporting PD patients and research in the United Kingdom) had also highlighted balance and walking problems as a top priority for future research. The strength of emerging evidence from initial testing was also interpreted as a clear rationale to embed the developed stimuli within new DHI. The general idea of using specific auditory cues to stimulate specific brain activities and promote improvement in balance and walking had also been presented several times at downstream stakeholder meetings (e.g., community-based Parkinson's UK branch meetings) along with descriptions of plans for future developments.

Once a prototype mobile application was developed, it was arranged for potential end-users to pilot the DHI and give feedback. This underlying concept for the DHI was born from the researchers' interpretation of scientific literature and their own laboratory-based observations. In discussions that followed piloting, user feedback can be summarized by the unanimous opinion that, while walking might show a noticeable improvement, listening to continuous sounds was not practical for PD patients in daily life due to the potential for both continued distraction and annoyance. In other words, the concept represented a compelling idea on paper, but the practicalities of integrating use of the DHI into daily life had simply not been sufficiently considered. Even though the background research and concept for the DHI was presented to stakeholders, the failure to actively engage these stakeholders in a meaningful process of evaluating the DHI within context of the lived experiences of PD patients, meant that the expertise held with the target users was not integrated at the earliest stages of DHI conceptualization.

An additional learning was that all prior presentations (i.e., engagement with end-users) reflected an 'us and them' approach, where the expert scientists were developing solutions *for* the patients, rather than *with* them; an approach that propagated an unhelpful hierarchy and actively discouraged patient stakeholders from offering insights, especially in the form of criticism, that would likely have helped avoid wasting resources. This serves as an example of researchers/developers attempting to engage with end-users, albeit too late and too little in the process.

11.3 Developing a New PPI Structure

In response to the experiences above, a new series of meetings with end-users were organized to present the project background and actively request comments and criticism. Patients with PD and FOG and their family members/carers were then invited to join a Project Advisory Group (PAG) with the aim of discussing the DHI in more detail. The initial group comprised of 5–7 members, and the initial intention of the PAG was to address our first recommended principle and define the problem or challenge from the perspective of end-users.

The revised emphasis on inviting comments and criticism resulted in lively debate, where new ideas were voiced regarding the actual unmet needs of stakeholders. Patients expressed having low interest in PD symptoms being delayed or slightly reduced by the auditory cues. Rather, their greater need related to the moment when they experienced freezing and were trying to take a step. It was these 'critical moments' of FOG that people described feeling most vulnerable and in need of a solution to improve their quality of life.

However, it is important to note that, at this time, the power to make decisions was retained by the researchers. As such, while discussions were fruitful, an unhelpful hierarchy was maintained at that time. Nevertheless, the open dialogue between the researchers and end-users combined with a willingness of the researchers to accept criticism, served to fundamentally shift the researchers' understanding of the perceived problem that needed to be addressed by the DHI—from the patient perspective. This presented a significant challenge to the researchers, as they began to understand the shortcomings of their previous approaches and realized the need to abandon ongoing projects to realign objectives with that of the end-users.

11.4 Reflections and Learnings for Future PPI

In this case study, the intention was not to provide details of specific activities carried out within the HCPD framework, as these are unlikely to translate across different applied contexts and programs [19]. Section 8 of this chapter could be used as a general guidance for planning such activities. The critical point to be made from this case study description, is that the activities and discussions all served to improve the understanding of the established downstream problem to inform the realignment of the planned research. This represents an attempt to break down the hierarchy that existed in terms of the power to make decisions. It was understood within the PAG that decisions made within that forum would be upheld and that WY was accountable in this regard. For example, initial PAG meetings were co-chaired by someone with PD. This enabled the role of the researchers to progress towards that of a facilitator, rather than an expert voice in the room.

Over time, the issue of generalizability of PAG decisions emerged. While PAG outcomes were pivotal in realigning the research agenda, the group of stakeholders were small and it was important to gather insights from a larger and more diverse group to further scrutinize decisions to date, and future plans. This approach maintained the newly found commitment to invite criticism and be agile in response. This prompted exploration and scoping exercises with larger audiences, comprised mainly of Parkinson's UK branch meetings as well as HCP/patient meetings to actively promote discussion.

Over time, this naturally emerged as an iterative process of summarizing and sharing established aims and ideas from the PAG to larger communities of stakeholders and then comments and insights were invited and interpreted by the researchers to feed back to the PAG (shown in Fig. 1). In turn, this has permitted more in-depth processes of both: (i) implementation and planning, and (ii) reflection and iteration. Reinforced through the delivery of agreed independent research projects, stakeholders see their decisions being actioned at each stage of the research process, thereby reinforcing their continued engagement.

Within these activities it became apparent that many attendees felt reluctant to speak in front of the whole group, but would enthusiastically share experiences and insights in more private discussions either in person at the end of group meetings or over a separate scheduled telephone/video call. These individual conversations were considered to be important to overcome barriers to voicing opinions in group settings. However, accommodating these additional exchanges placed substantial demands on researcher time which inevitably detracted from the time allocated to conventional research and development activities. This raised an important question: *How much end-user engagement is optimal?* The answer will invariably depend on the applied context of the work.

It is important to highlight that, while the activities in this case study eventually grew to a level that is not feasible for the majority of research projects, the critical learnings and fundamental realignment that occurred in this case study resulted primarily from a small number of meaningful conversations that just needed to

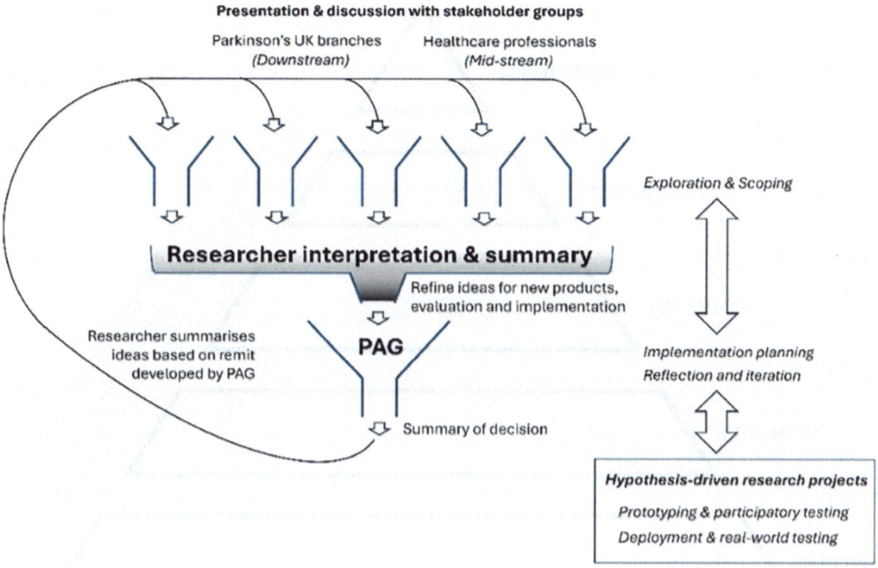

Fig. 1 PPI flow of communication between stakeholder groups

be approached sufficiently to identify major flaws in the researchers'/developers' perceptions. The key is for researchers to understand that there is a genuine need to learn and be agile in response to newly developed insights.

12 Conclusion

This chapter has emphasized that meaningful patient and participant involvement (PPI) is crucial for designing and developing effective digital health interventions (DHIs). The evolution from simple consultation to true partnership has revealed that inconsistent human-centered design often leads to mismatched goals, poor adoption, and wasted resources. A pragmatic, principle-driven approach to PPI ensures DHIs align with patient experiences, caregiver priorities, healthcare provider insights, and system stakeholder considerations, while also accounting for real-world implementation.

Effective stakeholder mapping and understanding their diverse needs and motivations, facilitates collaboration and prioritization. Integrating PPI early and throughout the innovation lifecycle helps define patient-relevant problems, select meaningful outcome measures, and anticipate implementation challenges. This enhances the feasibility, adoption, cost-effectiveness, and sustainability of DHIs, maximizing their impact on patient care and successful integration into health systems.

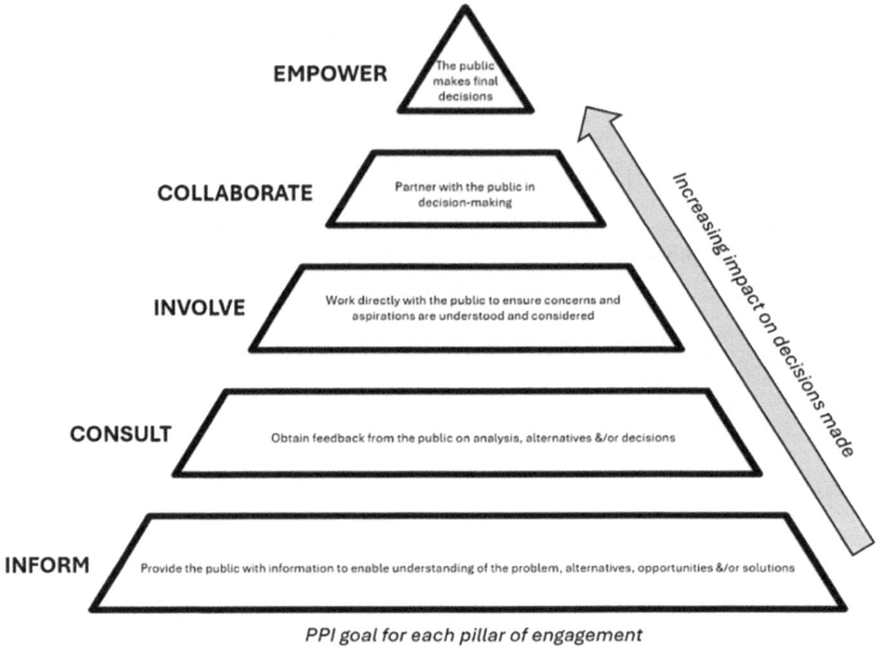

Fig. 2 Pillars of PPI engagement—adapted from the public participation pillars by IAP2

The case study on Parkinson's disease exemplifies how mismatched the good intentions of researchers and developers can be, with what patients prioritize as their unmet healthcare needs. It demonstrates the consequences of applying robust PPI practices later in the development process, and that it is possible to adjust course for better alignment with end-users. Accountability is key, to ensure that the feedback from PPI is considered and actioned where feasible (where it is not, communicate this back to stakeholders with a clear rationale).

As a final takeaway, we provide an adaptation of the Public Participation Pillars from the International Association for Public Participation (IAP2) to support an understanding of how researchers and developers new to PPI can understand the increasing pillars of engagement (shown in Fig. 2) [20]. We believe that starting at any point and working towards enhancing end-user engagement is a good approach to overcome the uncertainties and perceived lack of clarity around PPI.

Ultimately, PPI-driven design, going beyond superficial efforts, is a strategic imperative that accelerates the translation of evidence into practice. Fostering genuine co-design partnerships allows developers to create user-friendly, culturally sensitive, ethically sound, and scalable technologies. This improves patient outcomes and contributes to more efficient, equitable, and integrated healthcare delivery. Such an approach provides a roadmap for future DHI initiatives aiming for meaningful and lasting success.

References

1. M Bird M McGillion EM Chambers J Dix CJ Fajardo M Gilmour K Levesque A Lim S Mierdel C Ouellette AN Polanski 2021 A generative co-design framework for healthcare innovation: development and application of an end-user engagement framework Res. Involv. Engag. 7 1 12
2. Z Azizi JR Golbus EM Spaulding PH Hwang AL Ciminelli K Lacar MF Hernandez NA Gilotra N Din LC Brant R Au 2024 Challenge of optimizing medical therapy in heart failure: unlocking the potential of digital health and patient engagement J. Am. Heart Assoc. 13 2 e030952
3. Agali, K., Masrom, M., Abdul Rahim, F., Yahya, Y.: IoT-based remote monitoring system: a new era for patient engagement. Healthc. Technol. Lett. **11**(6), 437–446 (2024)
4. AC Morris S Douch T Popnikolova C McGinley F Matcham E Sonuga-Barke J Downs 2024 A framework for remotely enabled co-design with young people: its development and application with neurodiverse children and their caregivers Front. Psych. 15 1432620
5. AC Cole K Adapa A Khasawneh DR Richardson L Mazur 2022 Codesign approaches involving older adults in the development of electronic healthcare tools: a systematic review BMJ Open 12 7 e058390
6. RK Bajaj RM Meiring F Beltran 2023 Co-design, development, and evaluation of a health monitoring tool using smartwatch data: a proof-of-concept study Futur. Internet 15 3 111
7. M Ferrua E Minvielle A Fourcade B Lalloué C Sicotte M Palma Di O Mir 2020 How to design a remote patient monitoring system? A French case study BMC Health Serv. Res. 20 1 16
8. N Carey E Abathun R Maguire Y Wodaje C Royce N Ayers 2023 Co-design and prototype development of the 'Ayzot App': a mobile phone based remote monitoring system for palliative care Palliat. Med. 37 5 771 781
9. S Grosjean JL Ciocca A Gauthier-Beaupré E Poitras D Grimes T Mestre 2022 Co-designing a digital companion with people living with Parkinson's to support self-care in a personalized way: the eCARE-PD Study Digit. Health 8 20552076221081695
10. C Ferguson LD Hickman S Turkmani P Breen G Gargiulo SC Inglis 2021 "Wearables only work on patients that wear them": barriers and facilitators to the adoption of wearable cardiac monitoring technologies Cardiovasc. Digit. Health J. 2 2 137 147
11. N Isakadze N Molello Z MacFarlane Y Gao EM Spaulding YC Mensah FA Marvel S Khoury JE Marine ED Michos D Spragg 2022 The virtual inclusive digital health intervention design to promote health equity (iDesign) framework for atrial fibrillation: co-design and development study JMIR Hum. Factors 9 4 e38048
12. M Hanrahan C Wilson A Keogh S Barker L Rochester K Brittain J Lumsdon R McArdle 2025 How can patients shape digital medicine? A rapid review of patient and public involvement and engagement in the development of digital health technologies for neurological conditions Expert Rev. Pharmacoecon. Outcomes Res. 25 2 137 154
13. MDP Arias López BA Ong X Borrat Frigola AL Fernández RS Hicklent AJ Obeles AM Rocimo LA Celi 2023 Digital literacy as a new determinant of health: a scoping review PLOS Digit. Health 2 10 e0000279
14. E Aronoff-Spencer M McComsey MY Chih A Hubenko C Baker J Kim DK Ahern MC Gibbons JA Cafazzo P Nyakairu RC Vanderpool 2022 Designing a framework for remote cancer care through community co-design: participatory development study J. Med. Internet Res. 24 4 e29492
15. www.mayoclinic.org
16. WR Young MW Rodger CM Craig 2014 Auditory observation of stepping actions can cue both spatial and temporal components of gait in Parkinson's disease patients Neuropsychologia 57 140 153
17. MW Rodger WR Young CM Craig 2013 Synthesis of walking sounds for alleviating gait disturbances in Parkinson's disease IEEE Trans. Neural Syst. Rehabil. Eng. 22 3 543 548
18. WR Young L Shreve EJ Quinn C Craig H Bronte-Stewart 2016 Auditory cueing in Parkinson's patients with freezing of gait. What matters most: action-relevance or cue-continuity? Neuropsychologia 87 54 62

19. K Liabo K Boddy S Bortoli J Irvine H Boult M Fredlund N Joseph G Bjornstad C Morris 2020 Public involvement in health research: what does 'good'look like in practice? Res. Involv. Engag. 6 1 12
20. www.iap2.org

Pauline Frank has a multidisciplinary academic background with degrees in Pharmacy, Journalism, and Evidence-based Healthcare from Universities in Sydney and Oxford. She has worked extensively in community and hospital pharmacy settings as well as in the pharmaceutical industry. Her research contributions have focused on patient insights in oncology and patient engagement strategies in clinical trials, representativeness, and digital health.

William R. Young is an Associate Professor in Rehabilitation Psychology at the University of Exeter. With a background in biomechanics, psychology, and clinical science, his research investigates the role of anxiety and eye movements in gait disturbances, particularly in neurological conditions such as Multiple Sclerosis and Parkinson's. He is active in the development of low-cost digital health innovations and public engagement in health research.

Leveraging the Use of Wearables in the Workplace to Improve Healthcare Providers' Well-Being

Dimitrios Zikos and Kathleen Hodgkins

Abstract This chapter begins by examining the various stressors healthcare providers face in the workplace, shedding light on the physical, emotional, and mental challenges inherent in their roles. These stressors, including long hours, high-pressure environments, and the critical nature of patient care, can significantly impact their overall health and well-being. The chapter introduces the potential of wearable technologies as tools to help healthcare professionals monitor and manage their health in these demanding conditions. The reader will learn about the different types of functionalities that wearables offer and how they can address specific challenges faced by healthcare workers. For example, wearables can track stress levels, physical activity, sleep quality, and other health indicators, offering real-time feedback that enables early interventions and promotes self-care. Additionally, the chapter discusses how healthcare management can better leverage these devices to support staff wellness. It emphasizes the importance of fostering a culture that integrates wearable technologies into workplace wellness initiatives. Finally, the chapter introduces a systems approach, highlighting the interconnected roles of healthcare providers, patients, wearable technologies, and healthcare management. This approach underscores how leveraging wearables can not only improve the health of providers but also contribute to better care outcomes, creating a healthier and more efficient healthcare system.

1 The Nature of the Healthcare Profession

Healthcare is stressful. Professionals in this field often operate in demanding, high-pressure environments where they are expected to multitask and deliver clinical care with acute attention and exceptional skill [1]. Unlike most professions, the stress faced by healthcare workers is amplified by the critical and often life-altering

---

D. Zikos
Texas Tech University, Lubbock, TX, USA

K. Hodgkins (✉)
Central Michigan University, Mount Pleasant, MI, USA
e-mail: kate.hodgkins@cmich.edu

P. Eappen et al. (eds.), *Remote Monitoring and Wearable Devices in Healthcare*, Information Systems Engineering and Management 63,
https://doi.org/10.1007/978-3-031-98897-4_8

nature of their responsibilities [2]. Patient needs and expectations are unique and complex, requiring healthcare professionals to consistently make accurate decisions that directly impact lives. While these responsibilities are essential to patient outcomes, they can be overwhelming and have a profound effect on both the short- and long-term well-being of healthcare workers [3].

This high-stress environment frequently leads to burnout, emotional exhaustion, and physical strain. These issues not only affect healthcare providers' mental and physical health but can also compromise the quality of care they (3). The degree of stress varies across roles; for example, a primary care physician might face less acute stress compared to the fast-paced, high-stakes environment of an emergency room physician. Yet, regardless of the specific role, providing patient care is universally challenging due to the expectation of delivering sound guidance and decisions in critical situations [4].

Exacerbating this stress is the demanding nature of the healthcare workforce itself. Chronic understaffing, long hours, stringent time constraints, limited resources, and, in some cases, hazardous working conditions contribute significantly to the burden on healthcare professionals [5]. It is estimated that the U.S. spends nearly $15 billion annually because of burnout-related turnover in nursing and physician professions [5]. Proactively addressing these challenges is crucial to safeguarding the health and wellness of those who are responsible for caring for others [6].

2 Stress Factors and Implications in the Healthcare Environment

Healthcare needs and operations do not rest—there is always a need on a 24/7 basis. As previously indicated, staffing levels and limited resources have consistently been a concern for health professionals, but, more recently, the COVID-19 pandemic magnified these concerns on a local, national, and international level [5]. During the pandemic, clinicians were placed in contagious environments, putting themselves and their families at risk while infectious control personal protective equipment and supplies had international shortages [7]. Additionally, staffing shortages drastically increased due to isolation and quarantine measures when nursing and physician shortages were already present. These scary realities negatively impacted health systems because the clinical need was greater than what was available [5, 7]. Given the increase in healthcare demands throughout the pandemic, health professionals experienced a 249% increase in work-related illness and injury between 2019 and 2020 because of increased staffing shortages, constant crisis management, supply shortages, and even increased harassment [6]. While the pandemic amplified these challenges, healthcare professionals have consistently experienced higher rates of stress and burnout [5]. Now, even though the pandemic has been managed and appears more stabilized, healthcare professionals consistently experience job-related stress and a higher rate of burnout than non-healthcare professionals [8]. Burnout is defined as an occupational syndrome with significant emotional exhaustion, depersonalization, and low sense of personal accomplishment [5]. Research suggests many common-sense strategies exist for managing burnout, such as meditation, reading, or getting outside [9].

3 Wearables in Healthcare: More Than just a Patient Monitoring Tool

Technological advancement across the healthcare arena is more rapid than ever [10]. It is expected to continue to be an integral part of everyday life, as well as healthcare management for acute and chronic diseases [11]. Health providers across various settings and disciplines benefit from remote monitoring and wearable devices. For example, many individuals use smart devices that allow them to monitor their own health, such as physical activity, mood, sleep, oxygen levels, and heart rates in real-time, which can be shared with their primary care providers [12]. When under the care of a health provider, technologies can monitor chronic diseases more comprehensively and in real-time while remote [11]. As a result of these advancements, data collected is no longer a snapshot in time but a narrative over a longer period.

These advancements have become an essential part of medical management and prevention as the benefits include improved clinical trials, reduced healthcare costs, and better health control [10]. According to the CDC, in the U.S., there are approximately 129 million people with one major chronic illness, and over half of the population have at least two chronic diseases [13]. Since information is in real-time and more readily accessible, health professionals can respond more efficiently [10]. Response time to health warnings can be the difference between life and death. In emergency rooms or intensive care units, there is monitoring at the nurses' stations notifying providers if a patient's status changes drastically. Today, this type of notification can take place while the patient is at home and not just in the next room or down the hall.

In an environment where sensors are typically seen as tools for patient monitoring, they should also emphasize the need to expand this perspective by recognizing the value of wearables for healthcare providers. There is little research on the impact of wearables for health professionals specifically; however, wearables can enable health professionals to monitor their health, detect potential stressors early, and manage their well-being [14]. This, in turn, can lead to improved resilience, better stress management, and a reduction in burnout, ultimately contributing to safer, more effective patient care [14].

4 How Wearables Can Be Beneficial to Health Providers

In recent years, wearable devices have become central tools for patients to manage chronic conditions and provide a comprehensive picture of their physiological health to healthcare providers. These devices, including sensors for monitoring vital signs, such as heart rate, oxygen levels, blood glucose, and respiration, are increasingly used in both home and inpatient settings [15]. However, while these technologies benefit patient care, they also place additional responsibilities on healthcare providers, particularly nurses, who are tasked with monitoring these devices, ensuring they function correctly, and responding to alerts.

It is somewhat paradoxical that wearables and remote monitoring technologies, designed to enhance patient health and wellness, can simultaneously become sources of stress for healthcare professionals. While these tools aim to improve outcomes by providing real-time data and enabling proactive care, they often add to the workload of health professionals by requiring constant attention, data interpretation, and timely responses. The literature has recognized such challenges with continuous use of other technologies, such as with Electronic Health Records [16]. Furthermore, the expectation to seamlessly integrate these technologies into clinical workflows without sufficient training or support can amplify and exacerbate feelings of pressure and frustration [17]. This dual impact—enhancing care for patients while potentially burdening providers—highlights the need for careful implementation, adequate training, and thoughtful design of these technologies to ensure they support both patient care and provider well-being [17].

Wearables can play a significant role in improving the well-being of healthcare professionals by offering practical ways to monitor, analyze, and enhance both physical and mental health. These devices can track vital signs, such as heart rate, activity levels, and sleep quality, providing early warnings for stress, fatigue, or other potential health concerns [18].

Similarly, wearables can promote physical activity by setting movement goals, reminding staff to take short walks or even suggesting posture adjustments during long hours at a desk. Sleep hygiene is particularly important for healthcare workers, especially those who work irregular hours. Wearables can provide insights into sleep patterns and offer personalized recommendations to optimize rest, helping mitigate the effects of shift work [19].

On a broader level, these devices can foster team well-being by enabling group challenges, such as step competitions, to promote physical activity and camaraderie. Wearables also integrate seamlessly into wellness programs. By using the data these devices collect, healthcare systems can offer personalized health plans or provide incentives for meeting activity and wellness goals. Health insurance companies also offer wearable device incentives to promote wellness at the workplace; and in exchange for healthy behaviors, they provide to employers lower group premiums in their policy [20]. Additionally, wearables that track emotional states or allow mood journaling can help professionals recognize mental health concerns, with built-in links to counseling or support services [4].

Table 1 Wearable technologies for workplace wellness in healthcare

Technology functionality	Opportunity for health professionals	Healthcare administration support role
Track vital signs and other measures: heart rate, RPM, activity levels	Provide early warnings for stress, fatigue, or other health concerns	Personalized health plans or provide incentives for meeting activity and wellness goals
Provide real-time alerts for abnormal values/patterns	Help prompt timely interventions when unusual patterns are detected	
Devices equipped with stress-tracking features	Help identify when providers are under excessive strain	Identify and remove stressors where possible
Mindfulness tools or reminders for breathing exercises	Encourage relaxation during busy shifts	Provide time-out rooms in the workplace when possible
Sleep monitoring and sleep hygiene (particularly for those who work irregular hours)	Insights into sleep patterns and offer personalized recommendations to optimize rest, helping mitigate the effects of shift work	Promote an organizational culture where supervisors value provider well-being and prepare shifts that promote work-life balance Hiring decisions may contribute to "healthier" shifts
Monitoring physical activity	Maintain physical fitness which in turn may reduce stress	Step competitions to promote physical activity and motivate others
Track emotional states or allow mood journaling	Help professionals recognize mental health concerns	Provide and facilitate built-in links to counseling or support services
Aggregated data at an organizational level		Help identify systemic wellness issues, and guide workplace interventions

Education and training are essential to fully leverage these tools. Providing workshops and resources on interpreting wearable data can empower staff to make informed health decisions [21]. When thoughtfully implemented, wearables can help healthcare professionals achieve better health, reduce burnout, and enhance overall job satisfaction, all while contributing to a healthier, more resilient workforce. Table 1 summarizes the wearable technologies and functions that can be useful for health professionals, and the support role of healthcare administrators.

5 Existing Incentive Programs and Their Effectiveness

Many companies and organizations, not just health organizations, have turned to wellness programs or incentive programs to boost employee physical, mental, and emotional well-being in efforts to improve morale and engagement and enhance overall organizational culture and operations. Workplace wellness is defined as a supporting activity or policy designed to improve the culture and health outcomes within the organization [22]. Some examples of workplace wellness programs include a smoking cessation program, an employee assistance program, an onsite fitness center, or access to a nutrition counselor. While some companies implement these programs independently, other organizations provide this type of programming on a contractual basis. A few examples of these wellness organizations are Headspace, Virgin Pulse, and Health Advocate. By contracting with an organization that provides wellness programs to organizations, employees have ongoing access to various types of initiatives available to them, if desired. Moreover, some of these programs have financial incentives. For example, if an employee earns "points" by completing different wellness courses within the program, they could earn a certain amount of money or an item from a pre-identified list.

While there is a wide range of incentive programs, research studies have found a variety of effectiveness in these programs [23]. Wellness programs should be relevant to the health professions and can vary across disciplines and organizations [24]. Organizations, in general, benefit from well programs through improved absenteeism, productivity, healthcare expenses, culture, and overall happier employees [22]. Employees benefited through improved job satisfaction, health outcomes, health care utilization, employee relations and morale [22].

Examples of studies which emphasize the potential of incentive programs include the study of Reddy et al. [23] which found these programs can be effective for reducing absenteeism while improving morale and overall organizational culture found that these programs are effective for reducing absenteeism while improving morale and overall organizational culture. Additionally, Melnyk et al. found that patient care was improved, and medication errors were reduced by prioritizing their nurses' health. Specifically, when a nurse's health was supported in combination with shorter shifts, they were 3–9 times to have better mental/physical health, no/little stress, no burnout, and higher quality of life [25].

6 Commitment of Healthcare Management to Prioritize Using Incentive Programs and Their Employees' Health and Well-Being

Healthcare administrators need to recognize and address the wide range of challenges in the health arena, creating environments that prioritize the health and well-being of health professionals. According to the U.S. Centers for Disease Control and Prevention (CDC), hospital leaders can promote workforce wellbeing by communicating with staff, modeling wellbeing, and ensuring a safe work environment [13]. This goal can be achieved through designing healthy workflows and making thoughtful decisions regarding staffing that will not only enhance operational efficiency but also safeguard healthcare worker's well-being. These decisions include ensuring adequate staffing levels, implementing team-based workflows, providing easy access to patient information, and setting clear standards of practice (18).

Research indicates that health providers do not receive standard training on emotional management in medical training; however, those who have received some training are more comfortable with managing emotions, including their patients and their own [26]. Additionally, physicians' health and exercise habits positively impact their patient care since they understand and promote standard guidelines because of practicing them personally [27]. Physicians who are not experiencing burnout and who feel healthy and satisfied with their jobs are more likely to deliver high-quality, safe care. Conversely, stress and burnout are linked to an increased likelihood of errors, which can have significant and sometimes devastating consequences for patients. This underscores the importance of supporting the health and well-being of healthcare workers—not just for their benefit, but also for the safety and quality of patient care.

Similar to a physician's health impacting their patient care, administrators' health impacts employee or organizational care. Health management and leadership can serve, not only as policymakers, but also as a partner in employee health [28]. By modeling healthy behaviors and practices for and with their team, leadership can serve as role models and in turn, provide support and empowerment to their team [28]. It could be argued, then, that health administrators are pivotal in improving organizational health since they lead the environment and culture through the organizations' operations and policies. In turn, as studies have shown, the wellbeing of health providers improves the quality of care, productivity and patient satisfaction [29].

7 Wearables a Feedback Mechanism: From Monitoring to Actionable Strategies

Using a systems-level approach, health administrators can determine where wearables fit into the "ecosystem" of a healthy work environment. Wearables provide individuals with real-time information, such as exercise intensity and frequency, heart rate, step count, or noise environment, and they can issue notifications for potential concerns, such as elevated heart rates [18].

Despite wearables providing real-time information, healthcare providers must be willing to use the data to improve their health and well-being. Integrating wearables into an organization's "ecosystem" of a healthy work environment is just one step to gathering actionable information. However, information alone is not enough. While wearables can raise awareness, health providers must actively respond to feedback for the data to have a meaningful impact [1].

Health administrators must identify the best ways to motivate and empower healthcare providers to use wearable feedback to improve their health. The authors believe wearables used by healthcare professionals can also provide valuable feedback to incentivize organizations to improve staffing by highlighting the impact of workload on well-being and performance. As mentioned earlier, wearables track metrics, such as physical activity, stress levels, and sleep patterns, by offering objective data that reveals signs of overwork or burnout. Trends of high stress or inadequate recovery can prompt hospitals to address staffing shortages or redistribute workloads. Furthermore, this data quantifies the toll of inadequate staffing on health professionals, providing measurable evidence to justify systemic changes. Also empower professionals to advocate for improvements with objective data, compelling administrators, stakeholders, and other decision-makers to act. Hospitals can use this feedback to build a strong business case for hiring more staff, creating a healthier workforce, and improving patient care outcomes.

8 The Need to Leverage Wearables—An Open Forum in the Health Industry

Wearable devices offer health providers more than just tools for monitoring their own health and well-being; they also present an opportunity for healthcare systems to support and maintain a healthy, satisfied workforce. Recognizing their potential, it is essential to harness the capabilities of wearables beyond their traditional role as personal lifestyle aids for physicians and nurses, using them strategically to enhance workplace health initiatives and overall employee satisfaction [18].

Engaging the healthcare community to improve the wellness of providers using remote monitoring devices requires a collaborative approach. The starting point should be identifying key stakeholders, such as doctors, nurses, administrators, and wellness experts, to ensure diverse perspectives are represented. Hosting listening sessions or virtual meetings allows healthcare professionals to share their experiences and provide insights on how these devices can support their well-being [1]. These initiatives, driven by health administration and backed by leadership, require full commitment. This commitment must address barriers to wearables, including privacy and confidentiality, employee compliance, technology durability, and cost/benefit ratios [30]. Addressing barriers will be different for each organization depending on their location, employee demographics and socio-economic statuses; however, building trust by ensuring privacy and security of health information and presenting evidence of how remote monitoring has improved health outcomes in other settings may be important [30].

Concerns about the practical and real healthcare world challenges also need to be addressed. There is a reality in health systems, regarding deeply engrained and established organizational culture elements and these are not easily and always modifiable. It is therefore important to pair wearable data with structured organizational strategies and policy advocacy that acknowledge real-world constraints such as nursing shortages, limited resources, and political considerations. While wearables can highlight workforce stress and burnout, meaningful change requires, on top of leadership commitment, also transparent reporting and partnerships with labor organizations or policy advocates [31]. This would act as a catalyst to turn insights into action. Health workforce initiatives—such as value-based care models, and wellness incentives can create accountability and enable resource allocation aligned with workforce well-being goals [32].

Encouraging feedback through surveys or anonymous forums to understand concerns and expectations is also valuable. Positioning remote monitoring as part of a larger commitment to provider wellness, supported by leadership, can integrate wellness into organizational culture. Creating an environment where provider well-being is prioritized, maintaining open communication, and regularly reviewing progress to adapt plans as needed are key strategies. These steps may foster trust and encourage healthcare professionals to embrace remote monitoring as a valuable tool for improving their health and well-being.

9 Risks and Mitigation of Risks from Health Providers' Using Wearables

While wearables offer health insights, and can support providers' well-being, their use in the workplace could raise privacy, and regulatory concerns. It is possible that health workers may feel surveilled if they are not confident that their data is never going to be used to monitor performance or behavior. For example, it would be a reasonable concern for a health provider who experiences high levels of stress (evidently, because of their wearables data), that the management would flag him/her as high risk for patient safety and therefore make staffing decisions accordingly, aka the risk that data could influence work-related decisions such as promotions or layoffs [31]. This may add, to the provider, more stress, and mistrust, or unintended disciplinary actions. To address these concerns, clear policies must be established that define data ownership, consent, and limits on data use, especially regarding off-duty monitoring. These include and are not limited to establishing open for review regulatory frameworks and data protection measures tailored to employee privacy [16]. These measures should not only be transparent, but honest, and serve the purpose they were intended to.

10 A Conceptual Representation of Wearable Devices for Health Providers

As previously mentioned, hospital management must take a proactive role in supporting the health and well-being of their staff, including providing access to wearable technologies that can help them monitor and manage their health. By prioritizing the health of their employees, healthcare organizations can foster a more sustainable and effective care environment for both providers and patients. In a healthcare ecosystem where patients, providers, and management all coexist, the integration of wearable devices has the potential to create a more balanced, healthy, and safe environment for all stakeholders. It is crucial for hospital management to understand that ensuring the well-being of healthcare providers is just as important as protecting the safety and health of patients.

Figure 1 visually represents the interconnected roles of healthcare providers, patients, wearable technologies, and healthcare management in improving care outcomes. Patients are central to the healthcare system and receive improved outcomes due to advancements in wearable technologies and provider efforts. Providers are connected to patients directly and interact with wearable technologies to deliver better care. Their work is influenced by factors like management and controllable work conditions. Wearables serve as a key mediator between healthcare providers and the outcomes of care. They support both the improvement of patient safety and the reduction of provider burnout. Healthcare Management and Administration ensures that controllable factors—like adequate staffing, team-based workloads, and technology support—are optimized. Management directly influences wearable technologies and impacts provider satisfaction and burnout. Addressing these factors, in turn, helps reduce provider burnout and increases provider satisfaction, leading to better care outcomes. The goal to improve care outcomes and patient safety is achieved by integrating wearable technologies, optimizing controllable factors, and enhancing provider satisfaction. The flow of the diagram suggests a systemic approach where wearable technologies and management factors collectively enhance patient care and provider well-being.

Figure 1 which summarizes the ideas in this article, is grounded in well-known theories and is supported by research evidence. We will hereby outline three relevant theories to support the relevance or these associations: The Socio-Technical Systems Theory can explain the interaction between health providers, patients (who represent social systems), and wearables (which represent technical systems) and therefore their integration has the potential to improve health outcomes [33]. The widely used Donabedian Model can also support the connection between improvements in term of structure (e.g., staffing, administration), process (e.g., wearables themselves), and outcomes (e.g., patient safety) [34].

Finally, the Job Demands-Resources (JD-R) Model says that adequate staffing, technology support, and team-based practice can reduce burnout among providers [35]. In addition to these theories, several publications have shown that wearables enhance monitoring, disease management, and clinical decision-making [18] while staffing levels and support structures (which are administrative factors) are linked to improved patient outcomes and lower provider burnout [36].

11 Conclusion

In conclusion, wearables hold significant potential to help healthcare professionals manage their well-being, offering tools to monitor stress, physical activity, sleep, and other health indicators. These devices provide real-time feedback, empowering healthcare workers to recognize signs of burnout or physical strain and take proactive steps toward self-care. If health systems support and provide incentives to their providers to use wearables as part of their daily routines, health professionals will be able to have more control to manage the demanding nature of their work, and their employer will, in turn, benefit from a healthier workforce. By integrating wearables into their daily routines, health professionals can better manage the demanding nature of their work, ultimately improving both their personal health and their ability to deliver high-quality care.

However, it is essential to recognize that wearables are not standalone tools but components of a broader ecosystem. While they are widely used for patient monitoring, their value extends beyond this function. When embedded within a systems approach, wearables can play a critical role in enhancing the well-being of healthcare providers and fostering a healthier, more resilient workforce. This interconnected perspective emphasizes the roles of providers, patients, wearable technologies, and healthcare management in creating a supportive environment that prioritizes wellness.

Moving to the future, to have a better understanding of the quantities and qualities of the contribution of wearable devices to health workers' well-being, longitudinal studies and randomized controlled trials can measure changes in stress, burnout, and physical health over time. Survey-based cross-sectional studies could capture subjective well-being and provider perception regarding the effectiveness of wearables. Interviews and focus groups would be useful to discuss not only user experiences but also privacy concerns that we discussed earlier in this chapter. Examining organizational outcomes and conducting ethical impact assessments can also provide information about the elements of trust and policy, aka how data-use policies affect adoption, and in turn the overall effectiveness of the use of wearables.

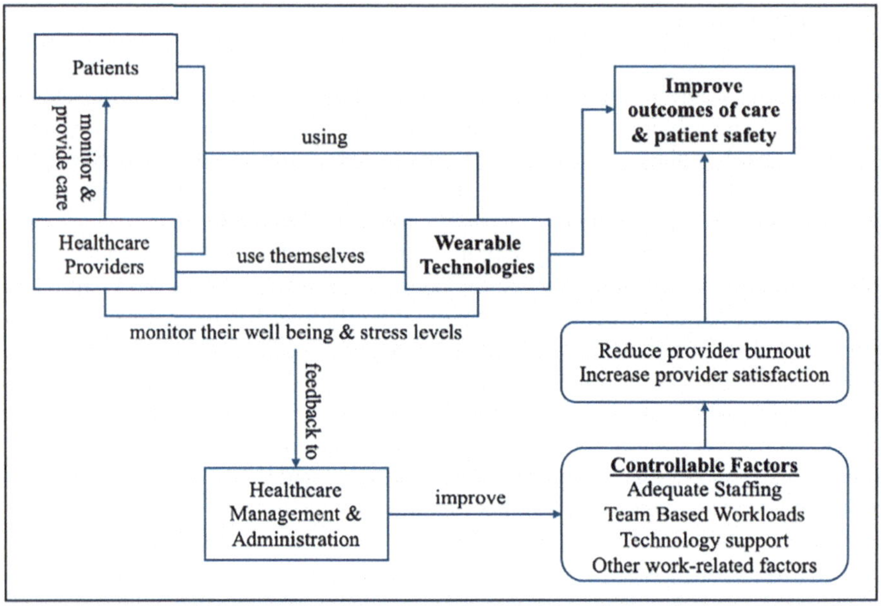

Fig. 1 Diagrammatic representation of a holistic, systems-thinking approach for wearables in healthcare

References

1. West, C.P., Dyrbye, L.N., Erwin, P.J., Shanafelt, T.D.: Interventions to prevent and reduce physician burnout: a systematic review and meta-analysis. Lancet **388**(10057), 2272–2281 (2016)
2. World Health Organization.: Health Workforce: Overview [Internet] (2020). Available from: https://www.who.int/health-topics/health-workforce#tab=tab_1
3. Shanafelt, T.D., Boone, S., Tan, L., Dyrbye, L.N., Sotile, W., Satele, D., et al.: Burnout and satisfaction with work-life balance among US physicians relative to the general US population. Arch. Intern. Med. **172**(18), 1377 (2012)
4. Upheal. Wearable Technology and Therapy [Internet].: Wearable Technology and Therapy (2024) [cited 2025 Apr 15]. Available from: https://www.upheal.io/blog/wearable-technology-and-therapy
5. U.S. Department of Health and Human Services.: Addressing Health Worker Burnout: The U.S. Surgeon General's Advisory on Building a Thriving Health Workforce [Internet], pp. 1–76 (2022). Available from: https://www.hhs.gov/sites/default/files/health-worker-wellbeing-advisory.pdf
6. Rotenstein, L.S., Brown, R., Sinsky, C., Linzer, M.: The association of work overload with burnout and intent to leave the job across the healthcare workforce during COVID-19. J. Gen. Intern. Med. **38**(8), 1920–1927 (2023)
7. Becene, I., Shah, A., Nguyen, K., West, M., Berrill, J., Stuart, J., et al.: In their own words: a qualitative survey of healthcare providers' experiences with personal protective equipment during the COVID-19 pandemic. Ann. Work Expo. Health **68**(5), 535–549 (2024)
8. Ungur, A.P., Bârsan, M., Socaciu, A.I., Râjnoveanu, A.G., Ionuţ, R., Goia, L., et al.: A narrative review of burnout syndrome in medical personnel. Diagnostics **14**(17), 1971 (2024)

9. Menardo, E., Di Marco, D., Ramos, S., Brondino, M., Arenas, A., Costa, P., et al.: Nature and mindfulness to cope with work-related stress: a narrative review. Int. J. Environ. Res. Public Health **19**(10), 5948 (2022)
10. Junaid, S.B., Imam, A.A., Balogun, A.O., De Silva, L.C., Surakat, Y.A., Kumar, G., et al.: Recent advancements in emerging technologies for healthcare management systems: a survey. Healthcare **10**(10), 1940 (2022)
11. Chen, Z., Liang, N., Zhang, H., Li, H., Yang, Y., Zong, X., et al.: Harnessing the power of clinical decision support systems: challenges and opportunities. Open Heart. **10**(2), e002432 (2023)
12. Fabbrizio, A., Fucarino, A., Cantoia, M., De Giorgio, A., Garrido, N.D., Iuliano, E., et al.: Smart devices for health and wellness applied to tele-exercise: an overview of new trends and technologies such as IoT and AI. Healthcare **11**(12), 1805 (2023)
13. Centers for Disease Control and Prevention. Professional Wellbeing [Internet] (2024). Available from: https://www.cdc.gov/niosh/healthcare/impactwellbeing/professional-wellbeing.html
14. Barac, M., Scaletty, S., Hassett, L.C., Stillwell, A., Croarkin, P.E., Chauhan, M., et al.: Wearable technologies for detecting burnout and well-being in health care professionals: scoping review. J. Med. Internet Res. **25**(26), e50253 (2024)
15. Hughes, A., Shandhi, M.M.H., Master, H., Dunn, J., Brittain, E.: Wearable devices in cardiovascular medicine. Circ. Res. **132**(5), 652–670 (2023)
16. Thomas Craig, K.J., Willis, V.C., Gruen, D., Rhee, K., Jackson, G.P.: The burden of the digital environment: a systematic review on organization-directed workplace interventions to mitigate physician burnout. J. Am. Med. Inform. Assoc. **28**(5), 985–997 (2021)
17. Hilty, D.M., Armstrong, C.M., Smout, S.A., Crawford, A., Maheu, M.M., Drude, K.P., et al.: Findings and guidelines on provider technology, fatigue, and well-being: scoping review. J. Med. Internet Res. **24**(5), e34451 (2022)
18. Piwek, L., Ellis, D.A., Andrews, S., Joinson, A.: The rise of consumer health wearables: promises and barriers. PLOS Med. **13**(2), e1001953 (2016)
19. Shin, S., Kim, S.H., Jeon, B.: Objective assessment of sleep patterns among night-shift workers: a scoping review. Int. J. Environ. Res. Public Health **18**(24), 13236 (2021)
20. Soliño-Fernandez, D., Ding, A., Bayro-Kaiser, E., Ding, E.L.: Willingness to adopt wearable devices with behavioral and economic incentives by health insurance wellness programs: results of a US cross-sectional survey with multiple consumer health vignettes. BMC Public Health **19**(1), 1649 (2019)
21. Ward, M.P., Malloy, J.S., Kannmacher, C., Steinhubl, S.R.: Educating the healthcare workforce of the future: lessons learned from the development and implementation of a 'Wearables in Healthcare' course. NPJ Digit. Med. **6**(1), 214 (2023)
22. Yadav, M.: A study of corporate wellness program as part of increasing employee productivity and its importance in post COVID-19 era. J. Sci. Technol. **6**(1), 488–493 (2021)
23. Jyothirmaye Reddy, S., Kalkar, P., Singh, A.P., Sharma, M., Sharma, P.: Investigate the effectiveness of employee wellness programs in reducing absenteeism, improving morale, and enhancing overall organizational culture. J. Inform. Educ. Res. (2024) [cited 2025 Apr 14]. Available from: http://jier.org/index.php/journal/article/view/889
24. Gerrard, K.M., Shepherd, J., Sehdev, R., Gangathimmaiah, V., Nagle, C.: Meeting the wellness needs of emergency department clinicians: a scoping review of interventions. Australas Emerg. Care **26**(2), 169–178 (2023)
25. Melnyk, B.M., Hsieh, A.P., Tan, A., Teall, A.M., Weberg, D., Jun, J., et al.: Associations among nurses' mental/physical health, lifestyle behaviors, shift length, and workplace wellness support during COVID-19: important implications for health care systems. Nurs. Adm. Q. **46**(1), 5–18 (2022)
26. Schwartz, R., Osterberg, L.G., Hall, J.A.: Physicians, emotion, and the clinical encounter: a survey of physicians' experiences. Patient Educ. Couns. **105**(7), 2299–2306 (2022)
27. Lampsas, S., Marinos, G., Lamprinos, D., Theofilis, P., Zakynthinos, G.E., Gialamas, I., et al.: Wearable activity trackers and physical activity levels among members of the athens medical association in Greece. J. Cardiovasc. Dev. Dis. **11**(10), 336 (2024)

28. Koinig, I., Diehl, S.: Healthy leadership and workplace health promotion as a pre-requisite for organizational health. Int. J. Environ. Res. Public Health **18**(17), 9260 (2021)
29. Sigurdsson, E.L.: The wellbeing of health care workers. Scand. J. Prim. Health Care **39**(4), 389–390 (2021)
30. Schall, M.C., Sesek, R.F., Cavuoto, L.A.: Barriers to the adoption of wearable sensors in the workplace: a survey of occupational safety and health professionals. Hum. Factors J. Hum. Factors Ergon. Soc. **60**(3), 351–362 (2018)
31. Bernhardt, A., Kresge, L., Suleiman, R.: Data and Algorithms at Work: The Case for Worker Technology Rights. [Internet]. UC Berkeley Labor Center, p. 46 (2021) [cited 2025 Apr 15]. Available from: https://laborcenter.berkeley.edu/wp-content/uploads/2021/11/Data-and-Algorithms-at-Work.pdf
32. West, T.H.R., Daher, P., Dawson, J.F., Lyubovnikova, J., Buttigieg, S.C., West, M.A.: The relationship between leader support, staff influence over decision making, work pressure and patient satisfaction: a cross-sectional analysis of NHS datasets in England. BMJ Open **12**(2), e052778 (2022)
33. Baxter, G., Sommerville, I.: Socio-technical systems: from design methods to systems engineering. Interact. Comput. **23**(1), 4–17 (2011)
34. Donabedian, A.: The quality of care: how can it be assessed? JAMA **260**(12), 1743 (1988)
35. Bakker, A.B., Demerouti, E.: The job demands-resources model: state of the art. J. Manage. Psychol. **22**(3), 309–328 (2007)
36. Aiken, L.H., Sermeus, W., Van Den Heede, K., Sloane, D.M., Busse, R., McKee, M., et al.: Patient safety, satisfaction, and quality of hospital care: cross sectional surveys of nurses and patients in 12 countries in Europe and the United States. BMJ **344**(mar20 2), e1717–e1717 (2012)

Dimitrios Zikos is an Associate Professor of Health Informatics at Texas Tech University Health Sciences Center. He holds a Ph.D. in Health Informatics from the University of Athens and specializes in data-driven decision-making, health analytics, and medical education technology. With a background that includes roles at Central Michigan University and the University of Texas, Dr. Zikos has contributed to over 100 publications and multiple EU-funded research projects.

Kathleen Hodgkins is Director of Clinical Services at Central Michigan University's Carls Center. She holds a Master's in Social Work from Michigan State University and is pursuing a Doctorate in Health Administration. Her focus is on healthcare professionals' work-life integration. She has worked extensively with interprofessional healthcare teams across clinical domains and is deeply involved in student training and healthcare system improvement.

Advancing Chronic Disease Management Through Wearable Technology: Opportunities, Challenges, and Future Directions

Gana Manjusha Kondepudi, Vijay Srinivas Pothula, and Kanikella Sivaji

Abstract Non-communicable diseases (NCDs), or chronic illnesses, cause approximately 41 million deaths each year, representing 71% of global mortality. Of these deaths, more than 15 million are premature, occurring between ages 30 and 69, with 85% happening in low- and middle-income countries. In India, NCDs are projected to cause 53% of all deaths, with estimated economic losses reaching $237 billion in the coming years due to early deaths from heart disease, stroke, and diabetes. Addressing chronic disease requires a multifaceted strategy focused on prevention and effective management. Preventive measures, such as diet, exercise, and avoiding tobacco, could prevent 80% of premature heart disease, stroke, and diabetes cases, and 40% of cancers. Wearable devices have become pivotal in chronic disease management, offering real-time health data, enhancing patient engagement, and supporting lifestyle changes. Devices like smartwatches, fitness trackers, and biosensors enable continuous monitoring, providing essential insights for personalized care. Studies show that integrating wearable technology into chronic care management may reduce hospitalizations and improve symptoms. Collaboration among technology developers, healthcare providers, and patients will be essential to maximize these devices' benefits while ensuring data privacy and security.

Keywords Chronic disease management · Wearable devices · Remote patient monitoring · Personalized health care · Health data privacy · Real-time health monitoring

---

G. M. Kondepudi (✉)
Department of Pharmacognosy, Vignan Institute of Pharmaceutical Technology, Visakhapatnam, Andhra Pradesh, India
e-mail: manjusha.kondepudi.g@gmail.com

V. S. Pothula
Stira Pharma LLC, Fairfield, New Jersey, USA

K. Sivaji
Department of Pharmacology, AKRG College of Pharmacy, Nalljerla, Andhra Pradesh, India

P. Eappen et al. (eds.), *Remote Monitoring and Wearable Devices in Healthcare*, Information Systems Engineering and Management 63,
https://doi.org/10.1007/978-3-031-98897-4_9

1 Introduction

1.1 Overview of Chronic Disease Management

This has made chronic diseases, such as cardiovascular conditions, cancer, diabetes, and respiratory disorders, urgent public health priorities because of the rising prevalence and the heavy burden they place on individuals and healthcare systems. Most prolonged health conditions are related to persistent health limitations and require continuing care and management. Such characteristics have made the rising prevalence of chronic diseases a result of demographic shifts and evolving health behaviors, thus making their management a key concern for public health initiatives [1].

Diagnosis of a chronic disease is such a life-changing event that tends to greatly alter the general wellbeing of an individual. Patients will probably undergo significant psychological and social changes in the initial months following a diagnosis as they attempt to come to terms with their illness. Every aspect of life is impacted by chronic illnesses, including daily routines, emotional fortitude, financial security, and physical health. The majority of patients encounter some level of activity limitation, lifestyle modifications, and ongoing medical care needs, all of which add up to the burden of disease. The patient's adjustment to the condition is made worse by the load, which consists of time, money, psychological issues, and emotional stress [2].

1.2 Psychological Adjustment and Well-Being

Onset of chronic disease is connected in a complicated way with subjective well-being. Hedonic adaptation theory predicts that individuals first experience a decline in happiness or well-being following a major life event, such as a diagnosis of chronic disease. Then they are predicted to gradually adapt and finally return to their baseline level of happiness. By contrast, the adjustment-to-chronic-conditions model posits that the trajectory of wellbeing post-diagnosis is highly variable: whereas some people successfully adapt, others may go on to struggle, resulting in persistently lower levels of wellbeing [3].

The empirical literature has reliably reported a decline in subjective wellbeing immediately following the onset of a chronic illness. Yet, the long-run consequences are mixed. Some studies find that people's levels of wellbeing return to those experienced before the diagnosis, implying good adaptation over time. In contrast, other studies find little recovery, and that people continue to experience lower wellbeing years after their diagnosis. A third group of studies shows uneven outcomes in which some groups show better adaptation than others with the influence of social support, access to health care and personal resilience as moderators.

1.3 The Role of Healthcare Access and Treatment Burden

Management of chronic diseases usually involves a high treatment burden. Patients must actively participate in different activities of self-management, such as taking their medications as prescribed, keeping and getting to medical appointments, and monitoring their health. These could be too many tasks, especially when coupled with poor access to health care resources. It would further exacerbate the treatment burden to those residing in areas where the health facilities lack high spatial access or have scarce service availability, and thus strengthen the negative effect on their wellbeing [4].

Among all the most significant determinants that can lower the load of chronic disease management, availability and access to health care services are of extreme importance. Access to health care providers, timely intervention, and availability of proper care systems can be a life savior for patients with chronic illnesses. These aspects not only make the physical and emotional weight of caring for a condition more bearable—they also enable patients to enjoy an improved quality of life. But there is still so much we don't know. Surprisingly, comparatively few studies have investigated how close or near a person is to health care services can affect their experience of living with a chronic illness and how it adds to their overall sense of well-being. It's a region that needs more focus and study to really know its potential impact on individuals' lives.

1.4 Need for Comprehensive Research

The immediate impacts are relatively well understood, but there is a general dearth of longitudinal analyses that take a comprehensive look at short-term and long-term trends. This kind of research is important to understand the complex influences of chronic diseases on people's lives. The identification of factors leading to successful adaptation or continued difficulty will enable health professionals and policymakers to design specific interventions that will improve the fate of people with chronic illnesses [5].

2 Understanding Wearable Technologies

Wearable technology has brought the best of health care into one stride and encapsulated innovation with accessibility in its revolution for the care of the patient and the management of disease. These wearable body-connected devices now contain the state-of-the-art sensors, wireless connectivity, and data analytics that converge to monitor many health metrics real-time [6]. Originally developed for fitness tracking, wearable technology has evolved into a very instrumental therapeutic tool in the management of chronic diseases, enabling preventive care and the development of highly personalized treatment plans. The broader category of wearable technology contains non-invasive devices that continually or intermittently capture an individual's physiological and behavioral data. Examples of such devices include smartwatches, fitness trackers, biosensors, and more advanced medical-grade devices such as continuous glucose monitors (CGMs) and electrocardiogram (ECG) sensors. Wearable technology has evolved from simple data collection to early event detection; real-time monitoring by these devices can be very personalized and can greatly engage people in the care of their health [7].

Functionality in wearable devices has increased manifold in the past few years. Older generation devices, pedometers, and basic heart rate monitors brought into focus the promotion of physical activities by way of tracking parameters including steps taken, distance moved, and calories burnt. Furthering this cause came the advanced wearable smartwatches and biosensors [8]. Nowadays, such wearable devices would track several parameters other than those probably mentioned in the question; to name a few, variability of heart rate, sleeping patterns, level of oxygen saturation in the blood, hydration, and indices for stress, to name but a few.

Wearables also provide an important place in the management of chronic diseases. For instance, continuous glucose monitors provide real-time information about blood sugar levels, hence allowing individuals with diabetes to make informed choices around diet, exercise, and medication. On a related note, ECG-enabled smartwatches can also detect abnormal heart rhythms—like atrial fibrillation—thus letting the user seek timely medical intervention. Hence, this innovation has shifted the paradigm of health care from reactive to proactive with great emphasis on prevention and early intervention [9].

This has been made even more useful with the integration of the technology into wearable devices combined with telemedicine platforms. This will bring remote monitoring capabilities that will enable the tracking by healthcare providers of patients' health metrics in real time, hence reducing the need for frequent hospital visits [10]. It adds to patient convenience and also helps in optimizing the resources of health care by focusing on early detection and preventive care. Now, with the aid of developments in AI and ML, wearables can process large data sets in the background to provide users with personalized insights and recommendations specific to their health profile [11].

Wearable technology is an agent of change in modern health care; it offers much more than just being a tool [12]. In doing so, wearables will continue to reformat the landscape into one that is all the more patient-centered and outcome-driven through constant engagement and tailored care. That is, the more the technology itself is developed, the greater its promise in finding answers to complex challenges within the realm of health and increasing the likelihood of desired patient outcomes related to chronic diseases [13, 14].

2.1 Important Features and Components of Wearable Devices

Wearable devices are advanced gadgets used in tracking, analyzing, and giving insights toward diverse health metrics. A wearable device is efficient since it combines both hardware and software; it tracks and collects data accurately and continuously while simultaneously analyzing.

Finally, as it involves critical biometric sensors, wearables in the form of photoplethysmography or PPG sensors, electrocardiogram or ECG monitors, and motion detectors are capable of playing a vital role in the measurement of vital signs, including heart rate, blood pressure, blood oxygen levels, and patterns of movements by which real-time monitoring of essential health parameters takes place [15–17]. Others would include a gyroscope and accelerometers to also quantify physical activity—something useful for tracking the body's odd, non-normative movements, like tremors or seizures.

Another core feature is that wireless connectivity can also enable this device to send out data onto a smartphone, cloud platform, or health system for further analysis. Technologies applied with regularity in integration, especially to devices, are Bluetooth and Wi-Fi, hence allowing remote monitoring and sharing of the same with providers of health services [18].

The wearable device itself is mostly composed of AI-driven algorithms and machine-learning models that process large datasets toward actionable insights and personalized recommendations. Indeed, in analyzing trends in health data, identifying anomalies, and even predicting possible complications such as glucose spikes and arrhythmias, algorithms are being used [19–21].

They also have very user-friendly interfaces, be it through a touchscreen, LED, or companion mobile app. They are comfortable to wear all day, being ergonomic, lightweight, and having a long battery life. This is further broadened in scope by the addition of environmental sensors to cutting-edge wearable technology for temperature, UV, and air quality tracking in preventative health.

This is where the management of chronic diseases has become a necessity, given these intertwined components—wearables—with increased user comfort, accuracy, and personalization [22].

2.2 Notable Examples of Chronic Illness Management

Wearable technology has brought new ideas in the domains of treatment and monitoring of chronic diseases. Few devices have stood out for application in the management of conditions like cardiovascular diseases, diabetes, neurological disorders, and obesity.

The Apple Watch and Fitbit also offer real-time monitoring of heart rate in the arena of cardiovascular health and can alert users to irregular heart rates, which may prove critical in the early diagnosis of diseases such as arrhythmias and heart failure

[23]. Advanced wearables with ECG sensors can detect atrial fibrillation, leading to early intervention by the doctor to prevent serious complications like strokes [24]. Also, portable blood pressure monitors such as the Omron HeartGuide will provide one with an opportunity to take a blood pressure reading at any time hence better manage hypertension [25].

These have revolutionized devices in the care of diabetes: Continuous Glucose Monitors, by using AI and machine learning in order to predict glucose spikes and dips, allowing for better control over glycemic levels and a reduction in the development of complications associated with diabetes [26].

There, wearables are used in the tracking of brain activity to help in the identification of seizure patterns with the aid of Empatica's Embrace Watch, thereby alerting carers at an early stage of such disorders in the case of neurological ones [27]. Wearables, by working through motion-sensing, help with the monitoring of Parkinson's patients; they have shown insight into the progression of symptoms through the tracking of changes in tremors and gait and how effective treatment is [28].

Wearable technology has also helped in the management of obesity; these devices support physical activity, track calories taken, and provide feedback to encourage healthy behavior. Quite a number of studies have shown that when such tools become part of life, tremendous improvement in weight management occurs, and a reduction in the complication associated with obesity is seen [29].

With the critical health data in collections, its analyses, and insights brought by means of wearable devices, which tend to mark a growing impact on technology intervention for chronic conditions while granting a way through for the patients toward better livelihood quality and enhanced health parameters [30].

3 Opportunities in Chronic Disease Management

3.1 Real-Time Monitoring of Health Metrics

The advent of technology has revolutionized health care, especially in the management of chronic diseases. Real-time monitoring makes it possible to keep a close eye on patients' health metrics continuously, yielding crucial data that can be used for timely interventions. It is now possible to measure important health parameters such as blood sugar levels, blood pressure, and oxygen saturation in real time using wearable devices like smartwatches, glucose monitors, and heart rate trackers [31].

The benefits of real-time monitoring include the possibility of healthcare providers detecting deviations from normal parameters and acting swiftly. For example, a sudden rise in blood pressure can be detected on the spot, and the health professional is able to change some aspects of the treatment plan or advice about lifestyle there and then. Second, this decreases the number of in-person visits and hence makes the healthcare service more accessible, relatively speaking, to patients with restricted mobility or located in faraway locations [32].

Another advantage of these devices, interoperability is achieved with the mobile applications and cloud platforms. Such technologies provide safe data exchange between patients and health professionals in taking care of disease management. Further, the latest developments in artificial intelligence have boosted the predictive ability of monitoring systems; hence showing potential health risks even before they become critical [33].

3.2 Early Detection and Prevention

Early detection and prevention may mitigate the progressive nature of chronic diseases like diabetes, cancer, and cardiovascular ailments. The screening programs powered by data analytics and sophisticated diagnostic tools are key components of this strategy [34].

Innovations in health care technologies include machine-learning algorithms that can analyze big data to predict the probability of a person developing a chronic condition. For example, AI-driven analyses of imaging results have been reported with remarkable accuracy in the detection of early-stage cancers. Similarly, predictive models can estimate the predisposition of an individual to diabetes based on his age, weight, and other genetic predispositions [35].

Prevention also includes lifestyle alteration. Tailored health counseling may be offered in real time on the data reached from wearable devices and, hence, guides individuals towards healthy behavior. More added educational campaigns, as well as community-based programs, are being implemented to help solidify an early detection along with a prevention regime [36].

3.3 Enhanced Patient Engagement and Self-management

Empowerment of the patient to take an active role in health is one of the fundamentals of efficient chronic disease management. More engagement would make the patient feel more ownership and responsibility; thus, more adherence to treatment plans leading to improved outcomes [37].

Digital platforms and mobile applications are spearheading this transformation. They have features that create medication reminders, diet and exercise trackers, and symptom diaries. In this way, the tools allow real-time feedback and actionable insight, enabling patients to stay on track with their health goals. These are further complemented by game-like features, like rewards for achieving milestones, thus enhancing user experience [38].

Telemedicine also helped in achieving the principle of self-management. Virtual consultations make it possible to consult health providers without necessarily visiting in person, hence providing uninterrupted care and support. The online access to educational resources as well as support groups allows patients to learn more about their conditions and connect with others facing similar challenges [39].

3.4 Data-Driven Decision Making for Clinicians

The integration of data analytics in health care has revolutionized the way clinicians think about chronic disease management. Data-driven decision-making enables the development of personalized care plans that address the unique needs of each patient [40].

EHRs represent an integrated platform that holds data on a patient, portraying their history in regard to health, test results, and the course of treatment taken. Information on such aspects gives the clinician an opportunity to make appropriate decisions in relation to changing the treatment status accordingly. Trends over time for patients' blood sugar levels can, for example, provide information to aid decision-making concerning changes in medication or diet recommendations [41].

Predictive analytics improves clinical decision-making. Through AI and machine learning, health professionals can now identify potential complications and take proactive action. For instance, predictive models can estimate hospital readmission risks; thus, allowing targeted interventions to prevent such readmissions [42].

Such data-sharing platforms also support collaboration among multidisciplinary teams. They ensure that all stakeholders, in this case, the primary care physician, the specialist, and the caregiver, have access to the same information in order to ensure coordinated care and better outcomes [43–46].

4 Technological Innovations

4.1 Sensors and IoT in Wearables

The convergence of wearable sensors and the Internet of Things (IoT) has had an impact on healthcare, particularly in the area of chronic disease management [47]. Such breakthroughs enable continuous monitoring, real-time data analysis, and personalised therapies, giving individuals and healthcare practitioners the information they need to make informed decisions [48]. The fast-growing wearable technology, combined with the frameworks of IoT, has a huge potential to improve the health outcomes and, more broadly, the quality of life for those with chronic conditions [49].

With IoT, the ecosystem of connected devices communicating and sharing information, such sensor arrays come together in order to make health monitoring effortless [49]. The data of the wearable sensors can also be uploaded onto cloud platforms where it will be processed and analyzed for real-time feedback and decision-making [50].

Application of wearable sensors has been utilized in chronic disease management by means of tracking vital signs, recognition of patterns, and showcasing trends in disease progression [51]. This is further supported by IoT, which provides the capability of remote monitoring required in facilitating communication between patients, caregivers, and healthcare professionals [52]. Some of the big challenges that are common to chronic disease management include early diagnosis, adherence to treatment, and prevention of complications [53].

4.2 *Wearable Sensors for Chronic Disease Management*

Wearable sensors can be classified in line with their functionalities and usage. The general types of sensors utilized in chronic diseases' management are elaborated upon as follows:

Physiological Sensors.

Physiological sensors relate to those used in the detection of physiological parameters as well as health indices, thus should be incorporated during treatment processes to cure diseases such as diabetes, cardiovascular disease, and respiratory disease [51]. Such examples include:

- Heart Rate Monitors: It monitors heart function and arrhythmias [51].
- Electrocardiograms (ECG): They are an indication that something is wrong with the functionality of the heart [51].
- Glucose Monitors: These are helpful to a diabetic patient because it lets them have a constant reading of glucose levels in blood [51].
- Pulse oximeters: They monitor the amount of oxygen in the blood, which is critical for patients with Chronic Obstructive Pulmonary Disease [51].

Motion Sensors.

Motion sensors, such as accelerometers and gyroscopes, are used to track physical activity and patterns [51]. They can also be applied in facilitating rehabilitation programs, detecting falls in the elderly and people with neuromuscular diseases [51].

Biochemical Sensors.

The biochemical sensors are designed in such a manner that they will detect biomarkers that include glucose, lactate and electrolytes in real time. Development of these sensors would place much emphasis on monitoring metabolic disorders and follow-ups in the responses of patients to different therapeutic interventions [52].

Environmental Sensors.

Environmental sensors detect parameters of the environment, including air quality, temperature, and humidity [53]. The sensors also are significant to patients with respiratory disease for supporting contextual health management [53].

Multisensor Systems.

Multisensor systems can combine various sensor types in a single device for multi-parametric monitoring [54]. In this regard, both smartwatches and fitness trackers are popular devices that embed physiological, motion, and environmental sensors [54].

4.3 IoT Frameworks in Wearable Technology

The IoT architecture in wearable technology consists of three layers of data collection, data transmission, and data analysis [55]. Each layer has a very important role in making the efficient management of chronic diseases possible [55].

Data Collection.

Wearable sensors can now continuously collect raw data for physiological, biomechanical, and environmental parameters [56]. That provides the basis for health insights—helping in the early detection of anomalies and monitoring disease progression [56].

Data Transfer.

IoTe-enabled wearables are using wireless communication technologies, including Bluetooth, Wi-Fi, and cellular networks, to transmit the data to either cloud platforms or edge devices [57]. In a way, seamless connectivity ensures that health providers can share and access the data in real time [56].

Data Analysis

Advanced data analytics techniques, machine learning, and artificial intelligence are used to convert raw sensor data into actionable insights [56]. These include personalized treatment plans, adherence monitoring, and predictive analytics [56].

4.4 Wearable Sensors and IoT Applications for Chronic Disease Management

4.4.1 Diabetes Management

CGMs, for instance, give real-time monitoring of blood glucose levels [54]. Data from this can be integrated into the IoT platforms for providing intelligent feedback to patients and healthcare providers [55]. Such a system, when integrated with AI algorithms, can serve in glucose trend prediction, suggesting diet changes, or even alerting users about possible hypoglycemic or hyperglycemic events [55].

4.4.2 Cardiovascular Disease Management

The wearable ECG sensors and heart rate monitors are useful in the study of arrhythmias, blood pressure monitoring, and generally in diagnosis related to cardiovascular diseases [48]. The IoT platforms offer valuable remote monitoring that results in early intervention with a reduction in the rehospitalization [55]. These have been used in the treatment of neurological illnesses such as Parkinson's disease and epilepsy [56].

4.4.3 Respiratory Disorders

Pulse oximeters and respiratory rate monitors are crucial for managing COPD and asthma [55]. IoT frameworks enable continuous monitoring of oxygen saturation and respiratory patterns, providing early warning of exacerbations [58]. Environmental sensors can detect air quality and allergens, assisting patients in avoiding triggers [55].

4.4.4 Neurological Disorders

Wearable sensors include accelerometers and gyroscopes. These have been applied in the treatment of neurology-related diseases like Parkinson's disease and epilepsy [55].

4.4.5 Chronic Pain and Rehabilitation

Wearable IoCT sensors also allow for the monitoring of physical activity, posture, and muscle engagement during rehabilitation programs [56]. One receives feedback on their execution of the activities and their recovery trend [56]. The continuum of monitoring oxygen saturation and respiratory patterns by IoT frameworks thus alerts about the exacerbations at an early stage [56]. The environmental sensors perceive air quality and allergens, helping the patient stay away from the triggers [56].

4.4.6 Monitoring Mental Health

For example, wearable sensors will measure physiological markers of stress, anxiety, or depression, including heart rate variability or skin conductance [57]. These metrics are then analyzed by the IoT solutions that offer the user one-on-one-based interventions in the form of mindfulness exercises or alerts for professional help when needed [57].

These devices track the rhythm of movements, monitor tremors, and detect gait [58]. IoT-based platforms update the caregiver in real-time, and hence increase the safety of patients and the efficiency of treatment [58].

4.4.7 Case Studies

In 2019, the IoT healthcare sector was valued at USD 113.75 billion and is expected to grow to USD 332.67 billion by 2027, at a compound annual growth rate of 13.20% [59], emphasizing the growing popularity of wearable technologies in healthcare owing to their potential for real-time monitoring. The global IoT market grew dramatically, reaching USD 100 billion in 2017 and projected to reach USD 1.6 trillion by 2025.IoT-enabled wearable devices are transforming healthcare by enabling remote monitoring and timely interventions as mentioned in Table 1.

4.4.8 Integrating Wearables with EHR: Transforming Healthcare with Real-Time Data

The integration of wearable devices with electronic health records alters the game in the health-care context. Wearables, using sensors, may now detect clinical and behavioral aspects such as heart rate, physical activity, and sleep in real time and send them to EHR systems. This fluid flow of data contributes to a holistic view of patient health, enabling early diagnosis of potential health problems and chronic condition management, while at the same time expanding possibilities for tailored care plans. In addition, real-time data empowers patients to make shared decisions and be actively involved in their care [73].

4.4.9 Clinical Implications of Wearable Integration

Several research programs have already demonstrated the therapeutic relevance of integrating wearable data into EHR systems by analyzing longitudinal Fitbit and EHR data to explore relationships between daily step counts and onset of illness across

Table 1 Table captions should be placed above the tables

Focus	Key findings	References
Eye drop usage tracking and fluid level monitoring	Developed a 3D-printed sleeve for eye droppers; achieved 97% accuracy in detecting usage	[60]
Wireless communication in wristwatch-based health monitoring	Designed a sensor platform using an 868 MHz Mi-Wi network; achieved 31 m communication range, outperforming fitbit charge 3	[61]
Flexible resistance strain sensors for motion and physiological monitoring	Prototypes made with carbon black and nanotubes show potential for human motion monitoring	[62]
Squat posture classification using IMUs and deep learning	Achieved 58.7% accuracy with IMU on the right thigh using conventional machine learning	[63]
Fall prediction using accelerometers	Fast and slow modes achieved sensitivity/specificity of 97.8%/99.1% and 98.6%/99.3%, respectively, validated with MobiFall dataset	[64]
Creatinine detection in serum	IoT-based point-of-care device detected creatinine up to 50 ppm, but human sample testing was not performed due to ethics issues	[65]
Monitoring skin temperature, heart rate, and body impact	Suggested improvements for IR sensors to reduce noise and improve usability for wrist-worn devices	[66]
Estimating fluid loss during physical activity	Developed sensors for monitoring heart rate, GSR, and skin temperature, transmitting data via ZigBee	[67]
Measuring serum CTx-1 concentration	Portable sensing device calibrated with known samples; capable of automatic urine sampling and cloud data transmission	[68]
Using wearable data to identify subtle infection indicators	Smartphone app collects smartwatch data, symptoms, and test results to identify COVID-19 signs	[69]
Examining security issues in IoT-based wearable systems	Proposed a set of security requirements and architectural considerations for wearable IoT systems	[70]
Security for implantable medical devices (IMDs)	Reduced computational overhead and message size for IMD authentication using fingerprint algorithms	[71]
ECG monitoring for nontechnical users	Proposed a cost-effective remote ECG monitoring system designed for long-term health tracking	[72]
Graphite/PDMS-based sensors for joint strain sensing	Successfully sensed strain on fingers, elbows, neck, and knees, showing potential for biomedical applications	[68]
Tracking medication adherence and intervention	Customized sensors successfully tracked eye drop medication adherence with Bluetooth-enabled communication	[60]
Improving accuracy in activity monitoring using IMUs	Compared deep learning with conventional machine learning for posture analysis; highlighted improvements in wearable motion tracking	[63]

the human phenome. The results presented evidence that wearables could generate meaningful insights for health practitioners to provide more targeted therapies. These applications epitomize how wearables can increase precision medicine and fill in the gaps in proactive healthcare management [74].

4.4.10 Barriers to Integration

Despite its potential, wearable integration into EHR systems is still in its nascent stages due to technical and privacy challenges. Most of the existing healthcare systems lack infrastructure for smooth interoperability and scalable workflows that assure data security. In addition, robust mechanisms are needed to secure patient consent, ensuring transparency in data usage to prevent misuse. The development of the EHR platform will have to make large longitudinal wearable data manageable in order to strike a balance between obtaining meaningful clinical insight and reducing false positives that can inundate the clinicians. [75, 76].

4.4.11 Improving Data Utility and Privacy

It would also be a big challenge to integrate wearable integration by protecting patient confidentiality and observing the policies of data protection. The FAIR principles—findability, accessibility, interoperability, and reusability—can make management of data easier. ML and AI algorithms will be at the forefront in deriving meaningful insights from wearable data and ensuring that only clinically relevant information reaches healthcare providers, thereby minimizing alert fatigue but maximizing the utility of wearables in decision-making [77].

4.4.12 Successful Strategies for Integration

An interdisciplinary approach is required for integrating wearables into clinical workflows. The first steps would be the development of educational resources for both patients and clinicians, identification of clinician champions who will advocate for wearable data integration, and assembling a technological support team. These champions should also address the benefits and limitations of wearables, so as to ensure realistic expectations and adaptability as technology evolves. The project "Integration of CGM Data into the HER" is just one of the many collaborative efforts that have produced a fertile framework through which to tackle the integration challenges. This 130+ member consortium with a diverse membership of stakeholders has produced pragmatic recommendations, implementation templates, and insight on how clinical workflows may be optimized for wearable technologies [78, 79].

5 Challenges in Implementation

The rapid development of wearable devices and associated technologies has revolutionized the health care sector, with special reference to the management of chronic diseases. Chronic conditions, such as diabetes, cardiovascular diseases, respiratory disorders, and arthritis, need continuous monitoring and timely intervention. Wearable technology allows for unexampled flexibility; that is, real-time tracking of health that enhances patient outcomes. The major challenges that lie ahead in the adoption of wearable devices for chronic disease management include data security, affordability, and compliance by patients. This chapter attempted to focus on such challenges and strategies to enhance the integration of wearable technology in chronic disease care [80]. Top Challenges in Wearable Technology for the Management of Chronic Disease are given below.

5.1 Data Privacy and Security Concerns

Ensuring the privacy and security of the patient's data within wearable technology.

Challenges:

- No weak encryption for resource-constrained devices.
- No standardized measures across platforms.

Strategies:

- Use secure end-to-end encryption of data.
- Harness the blockchain and other decentralized networks that enable the secure storage of information.

5.2 Accessibility and Affordability Barriers

The high costs of wearable devices and associated infrastructure necessary for their functioning are often a barrier to access.

Challenges:

- No Initial expenses of wearable device acquisition and integration.
- No Poor availability in remote or underserved areas.

Strategies:

- Public–private partnerships to off-set wearable device expenditures.
- Creation of low cost, open source, wearable technologies.

5.3 Technical Limitations and Issues of Accuracy

The technical capability of wearable devices should be augmented to achieve secure health monitoring.

Challenges:

- Precision of the sensors to recognize small changes in physiology is not good enough.
- No interoperability among devices of different manufacturers.

Strategies:

- No Global development of standards on calibration of the sensors and data sharing.
- No Investment in high-tech sensor technology and materials creation

5.4 Patient and Provider Resistance to Adoption

Both patients and healthcare providers resist wearable technology adoption.

Challenges:

- Skeptical view of the usefulness of wearable devices
- Fear of job loss for healthcare professionals

Strategies:

- No Raise public awareness about the benefits of wearables.
- Demonstrate how wearable technology will augment healthcare providers, not replace them.

6 Future Directions of Wearable Technology in Chronic Disease Management

Wearable technology has transformed health care, especially regarding the management of chronic illnesses. Diabetes, hypertension, cardiovascular problems, and respiratory illnesses, for example, account for a significant percentage of total mortality and morbidity due to their prevalence in today's world.

With increasing demands on health systems, wearable devices help in facilitating proactive monitoring and management of diseases through enabling real-time insight and personalized intervention. This paper argues for the future directions that wearable technology in chronic disease management can take, focusing on aspects related to technological advancement, policy, and regulatory frameworks, challenges that need to be surmounted so that its potential is harnessed fully.

6.1 Technological Advancements in Wearable Devices for Chronic Disease Management

The next generation of wearable technology will be focused on increased sensor capability, better analytics, and complete integration with health systems. The major locations where this will appear are:

Advanced Biometric Sensors.

Wearables increasingly feature clever sensors that can monitor vital parameters important in the management of chronic diseases.

Innovations here include:

1. CGM: It has enormous potential in diabetes care; biosensor-based wearables allow real-time glucose level monitoring and are evolving to become less invasive, with technologies such as microneedles and non-invasive optical sensors under development [81].
2. Cardiovascular Monitoring: Advanced wearables can track irregularities in heart rhythm, blood pressure, and arterial stiffness—very necessary in the management of hypertension and in the prevention of stroke [82].
3. Respiratory Health Sensors: Potential impactful wearables for health monitoring of asthma and chronic obstructive pulmonary disease (COPD) due to the measuring of oxygen saturation, breathing rate, and carbon dioxide level [83].

Integration with Artificial Intelligence.

The AI-driven algorithm is revolutionizing the way wearables data are analyzed and used. The integration with the AI will let wearables to;

- Predict the exacerbation of diseases by analyzing data from historical to real-time trends.
- Provide actionable insights as well as tailored recommendations for who the device wearer is.
- Early detection of complications through predictive analytics [84].

6.1.1 Intelligent Textiles and Non-invasive Wearables

The next generation of wearable devices will be so much more than wristbands and patches: the smart fabrics with integrated sensors, which will be seamlessly integrated into clothing to provide continuous and unobtrusive monitoring. This will increase user compliance and data accuracy [85].

6.1.2 Connectivity and Interoperability

With the coming of 5G and IoT, the connectivity of wearable devices will increase, enabling seamless data transmission to healthcare providers and integration with electronic health records, rendering real-time monitoring that allows for timely interventions in the management of chronic diseases.

6.2 Policy and Regulatory Framework for Wearable Technology

To ensure safety, efficacy, and ethical use, wearable technology must be developed and used in accordance with strict standards and regulations. The most important areas are.

6.2.1 Privacy and Security of Data

Wearable gadgets generate massive data pools containing sensitive health information. There is an onus on the policymakers to enforce strict laws with regard to the protection of user privacy. Regulations such as GDPR and HIPAA have recently set the benchmark.

- Encryption Standards: Devices must conform to high standards of encryption for data in transmission.
- User Consent: There should be clear policies on the use of data and mechanisms for obtaining user consent to the collection of their data [86].

6.2.2 Standardization and Certification

Wearable devices undergo standardization testing and certification by some credible bodies. Regulatory bodies like the FDA and CE start formulating regulations about wearables in relation to their performance and safety.

- Clinical Validation: Devices must demonstrate clinical accuracy with solid trials.
- Post-Market Surveillance: A follow-up is necessary after the launching of devices into the market, to be able to address the various problems that could become apparent [86].

6.2.3 Facilitating Integration into Healthcare Systems

Policy should ensure there is an incorporation of wearable technology into already existing healthcare infrastructure, and that is achievable by

- Interoperability standards allow data to be conveniently shared.
- Reimbursement policies concerning wearable devices to increase adoption rate.

6.2.4 Ethical Consideration

It will have to be answered, since wearable technology itself presents important ethical concerns, regarding bias in AI algorithms and equitable access. Those policymakers should incentivize diversification in clinical trials and further ensure that those devices are equally accessible for use among underserved populations.

7 Conclusion

The application of wearable technology in the management of chronic diseases has been one of the landmark events in modern health care history, hence filling in all facets of innovation, patient empowerment, and clinical advancement. By its very nature, chronic diseases are usually long-term, usually progressive, hence placing all global healthcare systems under strain. That is where wearable technology brings unlimited opportunities with the increasing digitization of health care in modern times.

7.1 Bridging Gaps in Chronic Disease Management

Wearables redefine chronic condition management by offering real-time monitoring that enables early detection and increases patient engagement. It will also make the wearer more aware of his or her status of health and the capability of taking part in chronic condition management through continually tracking health metrics. Examples of devices playing an empowering role in these patients include continuous glucose monitors, fitness trackers, and wearable ECG monitors.

Integration of EHR by these devices allows a smooth flow of data between patients and healthcare providers. All these come together to make this ecosystem interconnected, where actionable insights are readily available to the clinicians for data-driven decision-making. The wearable technology strengthens the provider-patient relationship and paves the way toward personalized, preventive, and predictive care.

7.2 Technological Innovations: Reaping Their Benefits

The wearable technology spine is comprised of new sensors, IoT capabilities, and Artificial Intelligence (AI) algorithms. These advancements improve the diagnostic and therapeutic capabilities of wearables. Advanced sensors capture granular health data; on the other hand, IoT networks support seamless sharing of this data on all platforms. The data is then analyzed by AI and machine learning to the point that it can track behaviors, predict progression, and provide health advice tailored to each individual.

AI-powered wearables, for example, can distinguish between normal and abnormal health patterns and alert patients and professionals in a timely manner.

It is especially effective in disorders like diabetes, hypertension, and cardiovascular disease, where early intervention can make a significant difference.

7.3 Overcoming Implementation Challenges

Still, chronic disease management has some challenges associated with the adoption of wearable technology. Perhaps the most highly cited concern has been data privacy and security: what the patients have to be afraid of in regard to their sharing of health-sensitive information. This, therefore, calls for strong cybersecurity protocol development that guarantees following the observance of existing laws on data protection to ensure confidence and compliance.

Another major barrier is the accessibility and affordability of wearable devices, especially in low-resource settings, where high costs restrict their use. Affordable access for underserved populations can only be accomplished through collaborative efforts between governments, healthcare providers, and manufacturers.

It also has to respond to technical challenges in issues of accuracy and device malfunction so that the goal of ultimate reliability is achieved. A big one of the major challenges is reluctance to adoption by both patients and physicians, hence requiring awareness campaigns and training programs to highlight the benefits of wearable technology.

7.4 Future Perspectives and Global Impact

Looking ahead, new developments in downsizing, bio-integrated sensors, and enhanced AI applications will propel wearable technology dramatically forward. More sophisticated user-friendly devices will further accelerate the adoption rates, while policy reforms will standardize their integration into healthcare systems.

The benefits brought by wearables will be equitably shared, in trying to close health gaps using subsidized programs and public–private partnerships, so that the adoption of wearables globally in managing chronic diseases translates into better population health outcomes, lesser spending in healthcare, and improvement in the quality of life of individuals.

Wearable technology, in the final analysis, is at the forefront of the digital health revolution and promises to bring a paradigm shift in dealing with the rising tide of chronic diseases. By embracing innovation and overcoming the challenges of implementation, wearables can be one of the biggest collaborators to redefine healthcare delivery in a much more efficient, inclusive, and patient-centric manner. Seam-less integration of wearable technology with the existing systems will be one of the most critical cornerstones toward a healthy and resilient future of the health ecosystem as it further evolves.

References

1. Stacherl, B., Sauzet, O.: Chronic disease onset and wellbeing development: longitudinal analysis and the role of healthcare access. Eur. J. Pub. Health **34**(1), 29–34 (2024)
2. Williams, K., Jackson, S.E., Beeken, R.J.: The impact of a cancer diagnosis on health and well-being: a prospective, population-based study. Psychooncology **25**, 626–632 (2016)
3. Kettlewell, N., Morris, R.W., Ho, N.: The differential impact of major life events on cognitive and affective wellbeing. SSM Popular Health **10**, 100533 (2020)
4. Debnar, C., Carrard, V., Morselli, D.: The longitudinal impact of a chronic physical health condition on subjective well-being. Health Psychol. **40**, 357–367 (2021)
5. Gallacher, K.I., May, C.R., Langhorne, P., Mair, F.S.: A conceptual model of treatment burden and patient capacity in stroke. BMC Fam. Pract. **19**, 9 (2018)
6. Khalid, S., Zahid, P., Zaheer, S., et al.: Empowering chronic disease management: the transformative role of wearable technology in personalized healthcare. Biomed. J. Sci. Technol. Res. **59**(2), MS.ID.009268 (2024)
7. Siddiqui, S.A., Zhang, Y., Lloret, J., Song, H., Obradovic, Z., et al.: Pain-free blood glucose monitoring using wearable sensors: recent advancements and future prospects. IEEE Rev. Biomed. Eng. **11**, 21–35 (2018)
8. Kong, S.Y., Cho, M.K.: Effects of continuous glucose monitoring on glycemic control in type 2 diabetes: a systematic review and meta-analysis. Healthcare **12**(5), 571 (2024)
9. Ding, X., Clifton, D., Ji, N., Lovell, N.H., Bonato, P., et al.: Wearable sensing and telehealth technology with potential applications in the coronavirus pandemic. IEEE Rev. Biomed. Eng. **14**, 48–70 (2020)
10. Guk, K., Han, G., Lim, J., Jeong, K., Kang, T., et al.: Evolution of wearable devices with real-time disease monitoring for personalized healthcare. Nanomaterials **9**(6), 813 (2019)
11. Wan, J., Al Awlaqi, M.A., Li, M., O'Grady, M., Gu, X., et al.: Wearable IoT enabled real-time health monitoring system. EURASIP J. Wirel. Commun. Netw. **1**, 1–10 (2018)
12. Mohammed, K.I., Zaidan, A.A., Zaidan, B.B., et al.: Real-time remote-health monitoring systems: a review on patient prioritization for multiple-chronic diseases, taxonomy analysis, concerns, and solution procedure. J. Med. Syst. **43**(7), 223 (2019)
13. Bhaltadak, V., Ghewade, B., Yelne, S.: A comprehensive review on advancements in wearable technologies: revolutionizing cardiovascular medicine. Cureus **16**(5), e61312 (2024)
14. Henriksen, A., Haugen Mikalsen, M., Woldaregay, A.Z., Muzny, M., Hartvigsen, G., et al.: Using fitness trackers and smartwatches to measure physical activity in research: analysis of consumer wrist-worn wearables. J. Med. Internet Res. **20**(3), e110 (2018)
15. Hickey, A.M., Freedson, P.S.: Utility of consumer physical activity trackers as an intervention tool in cardiovascular disease prevention and treatment. Prog. Cardiovasc. Dis. **58**(6), 613–619 (2016)
16. Bayoumy, K., Gaber, M., Elshafeey, A., et al.: Smart wearable devices in cardiovascular care: where we are and how to move forward. Nat. Rev. Cardiol. **18**(8), 581–599 (2021)
17. Bertagna, B., Dhargalkar, E.C.D.A., Paulvin, E.C.D.N., Mowery, M.: The top 9 blood pressure monitors of 2024, according to health experts.

18. Vettoretti, M., Cappon, G., Acciaroli, G., Facchinetti, A., Sparacino, G.: Continuous glucose monitoring: current use in diabetes management and possible future applications. J. Diab. Sci. Technol. **12**(5), 1064–1071 (2018)
19. Ossig, C., Antonini, A., Buhmann, C., Classen, J., Csoti, I.: Wearable sensor-based objective assessment of motor symptoms in Parkinson's disease. J. Neural Transm. **123**(1), 57–64 (2016)
20. Phillips, S.M., Cadmus Bertram, L., Rosenberg, D., Buman, M.P.: Wearable technology and physical activity in chronic disease: opportunities and challenges. Am. J. Prev. Med. **54**(1), 144–150 (2018)
21. Dobkin, B.H., Dorsch, A.: The promise of mHealth: daily activity monitoring and outcome assessments by wearable sensors. Neurorehabil. Neural Repair **25**(9), 788–798 (2011)
22. Brickwood, K.J., Watson, G., O'Brien, J., Williams, A.D.: Consumer-based wearable activity trackers increase physical activity participation: systematic review and meta-analysis. JMIR Mhealth Uhealth **7**(4), e11819 (2019)
23. Wang, W.H., Hsu, W.S.: Integrating artificial intelligence and wearable IoT systems in long-term care environments. Sensors **23**(13), 5913 (2023)
24. Alfian, G., Syafrudin, M., Anshari, M., Benes, F., Atmaji, F.T.D.: Blood glucose prediction model for type 1 diabetes based on artificial neural network with time-domain features. Biocybernetics Biomed. Eng. **40**(4), 1586–1599 (2020)
25. Marvasti, T.B., Gao, Y., Murray, K.R., Hershman, S., McIntosh, C.: Unlocking tomorrow's health care: expanding the clinical scope of wearables by applying artificial intelligence. Can. J. Cardiol. **40**(10), 1934–1945 (2024)
26. Arya, S.S., Dias, S.B., Jelinek, H.F., Hadjileontiadis, L.J., Pappa, A.M.: The convergence of traditional and digital biomarkers through AI-assisted biosensing: a new era in translational diagnostics? Biosens. Bioelectron. **235**, 115387 (2023)
27. Razzak, M.I., Imran, M., Xu, G.: Big data analytics for preventive medicine. Neural Comput. Appl. **32**(9), 4417–4451 (2020)
28. Xie, Y., Lu, L., Gao, F., He, S.J., Zhao, H.J.: Integration of artificial intelligence, blockchain, and wearable technology for chronic disease management: a new paradigm in smart healthcare. Curr. Med. Sci. **41**(6), 1123–1133 (2021)
29. Pettus, J., Edelman, S.V.: Recommendations for using real-time continuous glucose monitoring (rtCGM) data for insulin adjustments in type 1 diabetes. J. Diabetes Sci. Technol. **11**(1), 138–147 (2017)
30. Gupta, A., Deokar, A., Iyer, L., Sharda, R., Schrader, D.: Big data & analytics for societal impact: recent research and trends. Inf. Syst. Front. **20**, 185–194 (2018)
31. Munos, B., Baker, P.C., Bot, B.M., Crouthamel, M., de Vries, G.: Mobile health: the power of wearables, sensors, and apps to transform clinical trials. Ann. NY Acad. Sci. **1375**(1), 3–18 (2016)
32. Le Dinh, C., Chuang, R., Chokshi, S., Mann, D.: Wearable health technology and electronic health record integration: scoping review and future directions. JMIR Mhealth Uhealth **7**(9), e12861 (2019)
33. Wang, Y.C., Xu, X., Hajra, A., Apple, S., Kharawala, A.: Current advancement in diagnosing atrial fibrillation by utilizing wearable devices and artificial intelligence: a review study. Diagnostics **12**(3), 689 (2022)
34. Areia, C., Biggs, C., Santos, M., Thurley, N., Gerry, S.: The impact of wearable continuous vital sign monitoring on deterioration detection and clinical outcomes in hospitalized patients: a systematic review and meta-analysis. Crit. Care **25**(1), 351 (2021)
35. Ummels, D., Beekman, E., Theunissen, K., Braun, S., Beurskens, A.J.: Counting steps in activities of daily living in people with a chronic disease using nine commercially available fitness trackers: cross-sectional validity study. JMIR Mhealth Uhealth **6**(4), e70 (2018)
36. Wu, M., Luo, J.: Wearable technology applications in healthcare: a literature review. Online J. Nurs. Inform **23**(3) (2019)
37. Konstantinidis, D., Iliakis, P., Tatakis, F., Thomopoulos, K., Dimitriadis, K.: Wearable blood pressure measurement devices and new approaches in hypertension management: the digital era. J. Hum. Hypertens. **36**(11), 945–951 (2022)

38. Bobet, D., Shabir, G.: The future of chronic disease management: AI-driven healthcare and predictive analytics (2024)
39. Arefin, S.: Chronic disease management through an AI-powered application. J. Serv. Sci. Manag. **17**, 305–320 (2024)
40. Dankwa-Mullan, I., Rivo, M., Sepulveda, M., Park, Y., Snowdon, J.: Transforming diabetes care through artificial intelligence: the future is here. Popul. Health Manag. **22**(3), 229–242 (2019)
41. Kasula, B.Y.: Machine learning applications for early detection and intervention in chronic diseases. Int. Trans. Artif. Intell. **6**(6), 1–7 (2022)
42. Ahmed, R.A.A., Al-Bagoury, H.Y.H.E.: Artificial intelligence in healthcare enhancements in diagnosis, telemedicine, education, and resource management. J. Contemp. Healthc. Analytics **6**(12), 1–12 (2022)
43. Pandey, K., Agarwal, R.: Prognosticating congestive heart failure: AI and predictive analytics solutions. Innovations Comput. Sci. J. **10**(1), 1–8 (2024)
44. Agarwal, R.: Predictive analysis in health care systems using AI. Artif. Intell. Healthc. 117–131 (2022)
45. Arshi, O., Chaudhary, A., Singh, R.: Navigating the future of healthcare: AI-powered solutions, personalized treatment plans, and emerging trends in 2023. In: 2023 International Conference on Artificial Intelligence Innovation in Healthcare Industry (ICAIIHI), vol. 1, pp. 1–6. IEEE, Piscataway, NJ, USA (2023)
46. Ganesh, C.S., Royyala, S.K., Goud, K.V., Kethireddy, B., Rout, S.K.: Data-driven disease prediction and lifestyle monitoring system. In: 2024 1st International Conference on Cognitive Green Ubiquitous Computing (IC-CGU), pp. 1–8. IEEE (2024)
47. Mundnich, K., Booth, B.M., l'Hommedieu, M., Feng, T., Girault, B.: TILES-2018, a longitudinal physiologic and behavioral dataset of hospital workers. Sci. Data **7**, 354 (2020)
48. Yau, J.C., Girault, B., Feng, T., Mundnich, K., Nadarajan, A.: TILES-2019: a longitudinal physiologic and behavioral dataset of medical residents in an intensive care unit. Sci. Data **9**, 536 (2022)
49. Burghardt, K., Tavabi, N., Ferrara, E., Narayanan, S., Lerman, K.: Having a bad day? Detecting the impact of atypical events using wearable sensors. In: Proceedings of the 14th International Conference on Social Computing, Behavioral-Cultural Modeling and Prediction and Behavior Representation in Modeling and Simulation (SBP-BRiMS), pp. 257–267. Springer (2021)
50. Ramanujam, E., Perumal, T., Padmavathi, S.: Human activity recognition with smartphone and wearable sensors using deep learning techniques: a review. IEEE Sens. J. **21**, 13029–13040 (2021)
51. Tavabi, N., Hosseinmardi, H., Villatte, J.L., Abeliuk, A., Narayanan, S.: Learning behavioral representations from wearable sensors. In: Proceedings of the 13th International Conference on Social Computing, Behavioral-Cultural Modeling and Prediction and Behavior Representation in Modeling and Simulation (SBP-BRiMS), vol. 12268, p. 245. Springer Nature, Cham, Switzerland (2020)
52. Ji, S., Zheng, X., Wu, C.: HARGPT: are LLMs zero-shot human activity recognizers? arXiv preprint arXiv:2403.02727 (2024)
53. Kim, Y., Xu, X., McDuff, D., Breazeal, C., Park, H.W.: Health-LLM: Large Language Models for Health Prediction Via Wearable Sensor Data. arXiv preprint arXiv:2401.06866 (2024)
54. Wang, H., Zhao, J., Li, J., Tian, L., Tu, P.: Wearable sensor-based human activity recognition using hybrid deep learning techniques. Secur. Commun. Networks **2020**, 2132138 (2020)
55. McQuire, J., Watson, P., Wright, N., Hiden, H., Catt, M.: A data-efficient vision transformer for robust human activity recognition from spectrograms of wearable sensor data. In: Proceedings of the IEEE Statistical Signal Processing Workshop (SSP), pp. 364–368. IEEE, Hanoi, Vietnam (2023)
56. Englhardt, Z., Ma, C., Morris, M.E., Chang, C.C., Xu, X.O.: From classification to clinical insights: towards analyzing and reasoning about mobile and behavioral health data with large language models. Proc. ACM Interact., Mobile, Wearable Ubiquit. Technol. **8**, 1–25 (2024)

57. Suh, S., Rey, V.F., Lukowicz, P.: Tasked: transformer-based adversarial learning for human activity recognition using wearable sensors via self-knowledge distillation. Knowl.-Based Syst. **260**, 110143 (2023)
58. Ferrara, E.: Large language models for wearable sensor-based human activity recognition, health monitoring, and behavioral modeling: a survey of early trends, datasets, and challenges. Sensors **24**, 5045 (2024)
59. Telugunta, R., Choudhary, S.: Internet of Things in Healthcare Market. Available online: Allied Market Research. Accessed 17 Nov 2021
60. Payne, N., Gangwani, R., Barton, K., Sample, A.P., Cain, S.M., Burke, D.T., Newman-Casey, P.A., Shorter, K.A.: Medication adherence and liquid level tracking system for healthcare provider feedback. Sensors **20**, 2435 (2020)
61. Kumar, S., Buckley, J.L., Barton, J., Pigeon, M., Newberry, R., Rodencal, M., Hajzeraj, A., Hannon, T., Rogers, K., Casey, D.: A wristwatch-based wireless sensor platform for IoT health monitoring applications. Sensors **20**, 1675 (2020)
62. Zhang, H., Zhang, Z., Gao, N., Xiao, Y., Meng, Z., Li, Z.: Cost-effective wearable indoor localization and motion analysis via the integration of UWB and IMU. Sensors **20**, 344 (2020)
63. Lee, J., Joo, H., Lee, J., Chee, Y.: Automatic classification of squat posture using inertial sensors: deep learning approach. Sensors **20**, 361 (2020)
64. Saadeh, W., Butt, S.A., Altaf, M.A.B.: A patient-specific single sensor IoT-based wearable fall prediction and detection system. IEEE Trans. Neural Syst. Rehabil. Eng. **27**, 995–1003 (2019)
65. Prabhu, S.N., Gooneratne, C.P., Hoang, K.A., Mukhopadhyay, S.C.: IoT-associated impedimetric biosensing for point-of-care monitoring of kidney health. IEEE Sens. J. **21**, 14320–14329 (2021)
66. Malhi, K., Mukhopadhyay, S.C., Schnepper, J., Haefke, M., Ewald, H.: A Zigbee-based wearable physiological parameters monitoring system. IEEE Sens. J. **12**, 423–430 (2012)
67. Suryadevara, N.K., Mukhopadhyay, S.C., Barrack, L.: Towards a smart non-invasive fluid loss measurement system. J. Med. Syst. **39**, 38 (2015)
68. Nag, A., Afsarimanesh, N., Feng, S., Mukhopadhyay, S.C.: Strain induced graphite/PDMS sensors for biomedical applications. Sens. Actuators, A **271**, 257–269 (2018)
69. Quer, G., Radin, J.M., Gadaleta, M.: Wearable sensor data and self-reported symptoms for COVID-19 detection. Nat. Med. **27**, 73–77 (2021)
70. Pal, S., Hitchens, M., Rabehaja, T., Mukhopadhyay, S.: Security requirements for the internet of things: a systematic approach. Sensors **20**, 5897 (2020)
71. Zheng, G., Yang, W., Valli, C., Qiao, L., Shankaran, R., Orgun, M.A.: Finger-to-heart (F2H): authentication for wireless implantable medical devices. IEEE J. Biomed. Health Inform. **23**, 1546–1557 (2018)
72. Spanò, E., Pascoli, S.D., Iannaccone, G.: Low-power wearable ECG monitoring system for multiple-patient remote monitoring. IEEE Sens. J. **16**, 5452–5462 (2016)
73. Master, H., Annis, J., Huang, S., Beckman, J.A., Ratsimbazafy, F., Marginean, K., et al.: Association of step counts over time with the risk of chronic disease in the all of us research program. Nat. Med. **28**, 2301–2308 (2022)
74. Dinh-Le, C., Chuang, R., Chokshi, S., Mann, D.: Wearable health technology and electronic health record integration: scoping review and future directions. JMIR Mhealth Uhealth **7**, e12861 (2019)
75. Banerjee, S., Hemphill, T., Longstreet, P.: Wearable devices and healthcare: data sharing and privacy. Inf. Soc. **34**, 49–57 (2018)
76. Smuck, M., Odonkor, C.A., Wilt, J.K., Schmidt, N., Swiernik, M.A.: The emerging clinical role of wearables: factors for successful implementation in healthcare. NPJ. Digit. Med. **4**, 45 (2021)
77. Wilkinson, M.D., Dumontier, M., Aalbersberg, I.J.: The FAIR guiding principles for scientific data management and stewardship. Sci. Data **3**, 160018 (2016)
78. Baig, M.M., GholamHosseini, H., Moqeem, A.A., Mirza, F., Lindén, M.: A systematic review of wearable patient monitoring systems: current challenges and opportunities for clinical adoption. J. Med. Syst. **41**, 115 (2017)

79. Espinoza, J., Klonoff, D., Vidmar, A.: iCoDE Report: CGM-EHR Integration Standards and Recommendations. Diabetes Technology Society (2022)
80. Mukhopadhyay, S.C., Suryadevara, N.K., Nag, A.: Wearable sensors and systems in the IoT. Sensors **21**, 7880 (2021)
81. Yoo, J.H., Kim, J.H.: Advances in continuous glucose monitoring and integrated devices for management of diabetes with insulin-based therapy: improvement in glycemic control. Diab. Metab. J. **47**(1), 27–41 (2023)
82. Bayoumy, K., Gaber, M., Elshafeey, A.: Smart wearable devices in cardiovascular care: where we are and how to move forward. Nat. Rev. Cardiol. **18**, 581–599 (2021)
83. Gagnon, M.P., Ouellet, S., Attisso, E.: Wearable devices for supporting chronic disease self-management: scoping review. Interact. J. Med. Res. **13**, e55925 (2024)
84. Shajari, S., Kuruvinashetti, K., Komeili, A., Sundararaj, U.: The emergence of AI-based wearable sensors for digital health technology: a review. Sens. (Basel) **23**(23), 9498 (2023)
85. Meena, J.S., Choi, S.B., Jung, S.B., Kim, J.W.: Electronic textiles: new age of wearable technology for healthcare and fitness solutions. Mater. Today Bio. **19**, 100565 (2023)
86. Huhn, S., Axt, M., Gunga, H.C.: The impact of wearable technologies in health research: scoping review. JMIR MhealthUhealth **10**(1), e34384 (2022)

Gana Manjusha Kondepudi is an Associate Professor in Pharmaceutical Sciences with 13+ years of experience. Her research focuses on phytochemical and pharmacological screening of herbal drugs, formulation of novel herbal and cosmetic products, and use of natural excipients in drug formulations. She has published extensively and is known for her contributions to herbal and pharmaceutical sciences, including her work on bio-impedance and wearable healthcare devices. She actively mentors students and is engaged in scientific proofreading and manuscript development.

Vijay Srinivas Pothula serves as Assistant Manager in pharmaceutical formulation development with expertise in analytical method development, regulatory documentation (ANDA), and GMP compliance. He excels in the use of advanced instruments such as HPLC, UPLC, LC–MS/HRMS, and GC. His work ensures regulatory alignment with FDA, ICH, and GLP guidelines and includes leadership in method transfers, impurity profiling, and extractable/leachable studies.

Kanikella Sivaji An Associate Professor at A.K.R.G College of Pharmacy, Mr. Sivaji has seven years of experience in pharmaceutical education. He focuses on pharmacological studies involving animal models and has published 29 research articles in respected national and international journals. He is an editorial board member and reviewer for various scientific journals.

Healthcare on the Move: The Evolution and Future of Wearable Health Tech

Monirul Islam

Abstract The paradigm shift of wearable technology in healthcare delivery is replacing the traditional wristband wearables with a timely, latest range of multi-functional, hand-free devices that are directly built into our bodies. This chapter describes the emerging landscape of whole-body wearable health tech. It discusses various innovations, from intelligent clothing to skin patches with embedded biosensors (and even implantable devices), all of which can be used for real-time monitoring, early diagnosis, and personalized intervention. In addition to the challenges of data privacy, integration with existing healthcare infrastructure, and ensuring equitable access, further chapter sections analyze the potential for wearables in chronic disease management (CDM), preventive care, and telemedicine. These developments can lower costs, improve outcomes, and change how healthcare practitioners practice medicine in a consumer-payer ecosystem that values remote care delivery with patient-centricity; we see implications for these trends across multiple healthcare areas. It is great to see a role for these in the future and where innovation will go. However, most of these devices were not FDA (or EU) approved and calibrated, indicating a volatile clinical outcome on a subject. The transition of wearable devices from their humble beginnings as simple fitness and activity trackers to sophisticated diagnostic and therapeutic medical devices marks a new age in which technology is not an adjunct but instead one of the main pillars of patient care. Wearables are where the future lies, and they have the potential to revolutionize healthcare as we could not even imagine before.

Keywords Wearable electronic devices · Preventive care · Health care wearables

M. Islam (✉)
Associate Professor, Department of Management, Sister Nivedita University, Kolkata, India
e-mail: moni05081985@gmail.com

P. Eappen et al. (eds.), *Remote Monitoring and Wearable Devices in Healthcare*,
Information Systems Engineering and Management 63,
https://doi.org/10.1007/978-3-031-98897-4_10

1 Introduction

Wearable technology is an electrical device worn on the body (i.e. smartwatches, fitness trackers and bright clothing). Sensors and the rest of those devices gather data on the user concerning physical activity and vitals, to name a few [1]. Wearable technology can meaningfully impact healthcare by allowing for real-time monitoring of patients, early detection of health issues, and engagement in the care of patients [2]. Wearables also aid clinicians in developing patient-specific treatment plans, increasing patient outcomes and lowering healthcare. Medical treatment is used for many health conditions and not only chronic illness, sleep disorders, and mental health diseases by way of wearables. This type of data monitoring permits healthcare professionals to monitor blood glucose levels, heart rate and other significant markers and modulate medication dosages and treatment regimens in real time via wearable devices [3]. Wearable device adoption has considerably increased in the last year or so. Wearable devices that will be used in health care have a bright future, as the market is poised for tremendous growth and devices are becoming more robust [4]. Wearable devices and the connected health space are changing healthcare to improve patient service and care delivery systems, quality, and efficiency. The evolving ecosystem will have significant repercussions on the entire industry. According to a report by Deloitte Insights, the global medical wearable devices market is expected to jump from around 275 million in 2021 to close to 440 million units in 2024. The global wearables and healthcare devices market is already moving, expected to reach 115.8 billion dollars in 2021 to over 380.5 billion dollars by the end of the forecast period 2028 [5]. Predictive healthcare uses AI and data analytics in real-time to discover impending health problems such as a heart attack or allergic reaction. Generate incentives for personal care wearables as they allow healthcare patients to be involved in real-time healthcare decisions. It assists in cost-efficient remote monitoring of patients, particularly in crucial parts of cancer care [6]. Also, 5G will enable wearables to be connected faster and reliably for real-time data transfer, pushing the boundary of enhanced connectivity. Bright clothing integrates health monitoring capabilities into everyday objects, and devices can include smart clothes. Better health with implants, such as neural interfaces, will be the next great frontier in healthcare.

The benefits of fitness trackers have moved beyond tracking steps and sleep to advance functions to identify arrhythmias using a smartwatch, glucose monitoring in real-time and predicting seizures [7]. It blurs the border between consumer electronics and medical devices, setting a new stage for preventive healthcare, ongoing monitoring, and patient empowerment through personalized decisions based on holistic insights [8]. It is challenging to overcome data quality, patient security, cost and regulatory compliance [9]. This chapter details the health wearables types, the pros and cons of each, and what wearable devices in medicine will be in the future. Wearable technology in healthcare adopts diverse approaches. For example, devices with EKG (electrocardiogram) sensors can collect data on a patient's heart activity, which is essential for detecting cardiac abnormality [10]. Added to this, it also helps sustain chronic disease wearable tech devices. For example, glucose monitors of diabetics have been formed to allow constant tracking and timely action. These technologies power the participating individuals and help healthcare professionals make better decisions, improving patient outcomes remarkably [11]. Wearable health technology is changing the face of healthcare by giving rise to a preventive care paradigm.

Real-time data provides data-driven decision-making and dramatically improves patient outcomes [12]. For example, devices with EKG sensors in wearables take physiological signals from patients and can track heart activity, an essential role in contemporary healthcare [6]. Wearable health tech can solve many problems, but issues including but not limited to data security, accuracy, cost and regulation, amongst others, must be tackled. Tackling these challenges is relevant to the success and acceptance of wearable devices in the clinical arena.

1.1 Key Milestones in the Evolution of Wearables

Wearable technology has existed since the seventeenth century. The history of the Abacus ring has been around for quite a long time. Moreover, wearables will tell about the IoT, everything that assimilates hardware, software, and application development. It is a network in healthcare, electronic devices with cloud users to assimilate personal health data via mobile applications [13]. They are powered by micro-smart sensors in unassisted devices and are used to gauge physical hormones such as heart rate, blood pressure, and body temperature [14]. As a note, this information may be harvested by other medical Internet of Things devices such as audible or able to swallow. Wearables are available in various sizes, small enough to drape over wrist-sized patches but sufficient to support treating or even preventing chronic diseases. The estimated market value of IoT technology by 2025 could be as high as $6.2 trillion, of which a good bit of these nine-tenths have to be devices in healthcare ($2.5 trillion), sickness treatment and manufacturing ($2.3 trillion) [5]. Moreover, wearable technology that links healthcare to empower patients using real-time monitoring and data-driven decisions is the future of healthcare [6]. In the revolution of the medical field, wearable technology enables real-time monitoring of vital signs so that healthcare providers, based on the data, can act to prevent it. Wearable technology in healthcare looks bright, with new projections to include predictive wearables, smart textiles, and implantable sensors to bring forth a revolution in healthcare.

1.2 Evolution from Simple Fitness Trackers to Advanced Medical Devices

The transition from a basic step-counter to actual tools, equivalent to medical devices in the healthcare field, has been a big one. This far-future promise is looking less speculative with every new project like Neuralink by Elon Musk to treat and improve cognitive disorders. Wearable and the integration of AI/machine learning support predictive analytics at scale and data-led insights to enable early disease detection and preventive healthcare. The shift of healthcare from reactive healthcare to preventive healthcare. The hyper-growth of wearable and Internet of Things (IoT) devices allows us to monitor ourselves [13] constantly. Beyond fitness trackers, healthcare wearables have widespread applications for health-conscious folks and those managing chronic disease conditions. Personalized and efficient care is at the heart of effective diagnosis and data management for patient care via these devices. Wearable medical technology is a transforming reality not only at a theoretical level but one that is already starting to impact lives. Below are a few real-world applications: a Walker Tracker, a Step Tracking App, Blood Pressure Monitors, a Health Monitoring Platform, and a Medical Study App, to name a few. Wearing wearable technology in healthcare has many advantages [15]. Wearables, for example, are a simple, non-invasive way to monitor health information and track exercise, activity levels, and fitness. Wearable medical devices typically consist of other wearable equipment that enable doctors to prevent and diagnose whether the wearers have unhealthy diseases like heart disease, diabetes or sleep disorders. One of the most important features of wearable technology is that it can send instinct alarms and notifications to the wearer [16]. An example is when a smartwatch can inform the person that they need to take medication or have a slow/fast heart rate. Wearables are used for the safety of a wearer, such as triggering emergency services if the wearable senses they have fallen or are having a cardiac event [17]. Wearable technology is integral to healthcare, providing many benefits to providers and consumers alike. With wearables, the space continues to evolve, and healthcare providers and patients must remain abreast of all things wearable technology today.

Continuous glucose monitoring (CGM) has transformed diabetes management and is one of the driving forces behind this shift in Healthcare Monitoring Devices. This cutting-edge wearable medical technology offers real-time, instant, and continuous monitoring of glucose levels without the finger-pricks. Real-time glucose readings every 5 min. This device permits the monitoring of glucose levels in users [18]. If glucose levels get too high or low, the system pings the user to inform them they must act.

2 The Current State of Wearable Health Tech

Wearable technology has transformed the healthcare landscape to deliver real-time health data and healthy patient outcomes. Below are the three popular applications of wearable technology devices in healthcare.

2.1 Fitness and Wellness

Wearable tech has seen a massive increase in consumer use cases in the fitness and wellness space. Fitness trackers and smartwatches are the most common wearable devices to track physical activity and vital signs. They can assess heart rate, steps taken in a day, and calories burnt [19]. Bright clothing and shoes are just some examples of wearable smart devices in the field of fitness technology and sustainability.

2.2 Medical Treatment

Doctors are integrating wearable electronic devices into medical treatment regimens to improve patient outcomes. Medical wearable devices can detect and report numerous health conditions directly to healthcare providers. EGM monitors, insulin pumps, and smart inhalers [14]. These wearable technology types could be of more aid to patients with chronic diseases (diabetes, heart disease).

2.3 Clinical Trials and Research

Wearable devices can collect and analyze health information. Researchers can monitor patient vital signs and health data in real-time using wearable devices, enabling more precise results and reducing visits to medical facilities due to wearable technology through patient compliance [1]. Wearables (e.g. head-mounted displays and smart glasses) create a virtual digital world for patients in the context of augmented reality or even VR clinical trials.

3 Emerging Innovations in Wearable Devices

However, tech is constantly evolving, and newly updated products hit the market thanks to their availability to consumers. Products of all sorts are included here, consumer equipment bought at the hands of the masses and medical devices used by health care providers [20]. As we advance through 2025, the landscape of wearable health monitoring devices continues to evolve, thereby offering more refined and user-friendly alternatives for health and fitness freaky enthusiasts and medical professionals. These devices track crucial health metrics and provide valuable insights to improve lifestyle. Table 1 is a list of the most awaited wearable health monitoring devices that are anticipated to impact this year significantly.

3.1 Wearable Fitness Trackers

Fitness trackers use sophisticated sensors to give us key knowledge about our daily health and well-being, which thus constitutes a vital part of life. The simple fitness tracker app is superior to the mill pedometer, as this app provides a complete system to gauge physical activity and wellness. Wearables enable people to tally their health with accuracy to monitor heart rate or nerve pulse rate [21]. In addition, the wristbands govern physical movements, count steps, sleep stages, and calories burnt. Tiny devices with sensors are attached to a phone, and mobile app developers have developed easy solutions. Syncing with several electronic devices like fitness bands and smartwatches is easy. It promotes healthy competition worldwide, bringing a sense of community and motivation. Moreover, this lets organizations create proprietary challenges to encourage well-being in their community.

Table 1 List of latest available wearable devices in the market

Wearable devices	Application
Apple watch SE 3	ECG
Apple watch 10	Blood pressure, sleep apnea detection
Apple ring	Heart rate variability (HRV), sleep quality, stress levels, PPG
Dexcom Stelo	Continuous glucose monitoring (CGM)
Huawei watch GT 5	Heart rate monitoring and sleep analysis
Polar ignite 4	ECG, real-time heart health and blood oxygen monitoring
Fitbit charge 7	Continuous electrodermal activity (cEDA)
Oura ring 4	Temperature sensors, sleep tracking
Xiaomi Mi smart band 10	Continuous SpO2 monitoring
IdentifyHer's Peri	Track and manage perimenopause symptoms

Source Author compilation

3.2 *Smart Health Watches*

Smartwatches have been in the electronic devices vertical for ages now. However, it only started to possess actual value concerning health as recently as a few years ago. In addition to making calls and messages or sending voice commands, Smartwatches can also give information about certain strenuous activities, stress levels or anomalies in the user's heart. Apple Watch and other innovative health bands have transformed wearable tech [22]. These record physical activities, track vital signs and discern arrhythmias. It is not just a smartwatch but the total health of a machine which understands everything from heart rates to sleep. What truly sets it apart, however, is the continuous detection of atrial fibrillation (AFib), a condition with disorganized heartbeats that can result in strokes and heart failure [11]. These devices can sync with our smartphones, providing a handy and out-of-the-box view of our health. Last is a result of recent progress in optical-sensor, semiconductors and AI-based technologies. They all endowed the basis for photo plethysmography (PPG) to assess the volumetric changes in blood circulation [23]. Besides, it can accomplish remote patient monitoring to see in heart disease. If the device detects a fall, it also alerts emergency services if not responded to by a specific period [24]. The Apple Watch leverages robust wearable sensors to process pulse data and recognize irregular heart rhythms, a potential indicator of AFib. This FDA-approved application can provide valuable information about the heart by capturing real-time ECG data. The Apple Watch has done plenty to advance health monitoring but faces some regulatory challenges. Recently, Apple had to stop selling some models due to a patent infringement lawsuit. The problem was with the technology of smartwatches to sense blood oxygen levels. These changes highlight the ever-shifting landscape of health tech regulations and the importance of maintaining compliance.

3.3 *Wearable ECG Monitors*

The wearable electrocardiogram (ECG) monitor is a significant development of health wearables. These sensors also allow for continual heart rate monitoring, which supplies vital data for cardiac arrhythmia detection. These wearables enable continuous heart rate monitoring, which provides critical data on possible coronary heart abnormalities. People with cardiac issues can be lifesavers, providing real-time information about their heart health and peace of mind [25]. As much as 37% of internet users in the United States rely on wearable utilities to keep track of their heart status [5]. Wearable technology-assisted medical devices will help manage cardiac-related illnesses and are used to identify many recurrent arrhythmias, including atrial fibrillation (AF). Modern equipment for electrocardiography is not only effective in measuring the quality of the signal but also in identifying if there are any changes to the normal allele [26]. The data can then be forwarded to a physician for secondary analysis and diagnosis of the AF's symptomatic or asymptomatic nature.

3.4 Wearable Blood Pressure Monitors

Hypertension is well known for its ability to create morbidity and mortality around the world. Having a BP monitor integrated into wearables increases the chances of enhancing hypertension screening daily. With recent advancements enabling wearable technology, consumers can measure blood pressure in less time. However, this method of monitoring BP has not yet been fully developed and needs more investigation. This wrist-worn device uses oscillometric technology to provide a much easier, faster, and more accurate feature for reading blood pressure than earlier. It also measures physical activities and sleep, making it more comprehensive than other applications in giving health information. They also include a heart health coaching app that gives the user relevant information from data collected by the device. It shares data and monitors stress levels and aspects of heart health, giving a generalized overview of the user's health. It provides people with tips on their health, depending on their hereditary profile. It can synchronize with different wearables to check health day and night.

3.5 Biosensors and Implants

Biosensors employ portions such as antibodies or enzymes that identify biochemical levels in the body. Biosensors and Implants are best suited for continuous or invasive monitoring levels of all the medical wearables described. Such devices can track a number of health indicators, such as the level of glucose in diabetic patients or cardiac function in patients with heart disease [8]. Typically, it is wearable in form and design, most newer models are stickers that use advances in sensor technology and Artificial Intelligence algorithms to monitor key health stats without a subject having to remain immobile. Some biosensors give the patient and the clinician a way to quantify important parameters such as cardiac output, respiratory rates, weight, or even the volume of body fluids like blood and urine. Therefore, the likelihood of avoiding heart failure also increases [27]. Some valuable ways they are helpful include early disease detection, recommending early prevention and treatment for diseases, quick response, and cutting out costly hospitalized treatment.

3.6 New Product Developments

New products and newer models are being developed, including the Google Pixel Watch 2 with improved health features and the Apple Watch Series 9. Moreover, Philips has presented the Avalon CL Fetal and Maternal Pod and Patch, which provide continuous, non-interference monitoring, proving that wearable healthcare technologies are being improved at a very high pace. CGMs were recently cleared by the FDA for use with diabetes and other disorders [28]. Moreover, companies like Fitbit, which deals with wearable healthcare technology, embrace innovation in health-related products that include AI and enhanced tracking systems for their products.

4 Opportunities for Seamless Care Delivery

Indeed, it is evident that the future of health care will use wearable technology, especially in tele-monitoring patients. Investigators also revealed from a survey that sixty per cent of the providers thought that wearable technology in healthcare helps monitor patients. A Medical Study App is an application of medical research that aims to understand the frequency of fainting and its causes. They obtain data from participants who wear devices. It also helps compile relevant information on the episodes of fainting, which medical researchers consider crucial. Wearable technology can also show how patients and people all around the globe are doing physically in this increasingly connected world and with self-approach to health conditions. It also means that through monitoring vital signs, physical activity, and even mental health, the healthcare sector will be able to evaluate the condition of patients everything in real-time. It becomes easier to notify the patients and remind them of their appointments, medication time, and other symptoms or reports to fill in by the patients or other home-delivered solutions [29]. Solutions include multicast data live-streaming to help patients manage their diseases at home, reduce cross-infection, solve hospital occupancy rates, reduce healthcare costs, and enhance patients' health as much as possible.

Wearable technology enhances the patient experience because the patient becomes an active participant in the therapy. The patient receives benefits in tracking their state, remembering the tasks and treatments that should be done, and facilitating communication with healthcare professionals. It is important to believe in the potential role of wearable technology in treating illnesses [30]. Hence, research focuses on making patient monitoring devices available, affordable, and easy to use. Whether it is a fitness tracker, a smartwatch, or specialized medical sensors, the ultimate goal should remain: metamorphosing information technology to enhance health leader decision-making and healthcare outcomes.

5 Future Trends and Possibilities

5.1 Predictive Healthcare Through Real-Time Data Analysis

AI and 5G offer significant prospects for the wearable devices market since the former is a core technology for devices, and the latter is an important growth factor. These technologies facilitate the accurate transfer and analysis of information in real time and improve the relevance and uses of wearable devices in healthcare. First of all, the advancement of new battery technologies may create new opportunities since current limitations in terms of longevity or efficiency of the devices may be resolved. Wearable technology is one field with the highest prospect for revolutionizing the healthcare system [31]. Wearable technology in healthcare and throughout this industry's development remains on the edge of new trends and ideas. Below are areas and trends in wearable technology for health care and their possible implications for the industry poised for growth.

5.2 Remote Monitoring and Decentralized Trials

Remote Monitoring, in simpler words, refers to healthcare from a distance. One of the significant breakthroughs in the health sector's realization of wearable technology is the ability to monitor patients from a distance [32]. Healthcare practitioners have an opportunity to monitor distant patients' health, resulting in easy access to healthcare. The patient lives far from the hospital or cannot often visit the healthcare facilities [33]. Bright clothing in this area will likely bring tremendous changes and added value to healthcare by offering accurate and instant monitoring information. Medical wearables can also track the patients' status, do not require them to go to the hospital, and are equally effective in preventing readmissions. It can also help doctors interact and attend to patients more effectively, thus improving their ability to diagnose their ailments quickly and effectively. Second, worn technologies have the potential to contribute towards a decrease in health care expenses due to the timely identification of human problems, which eliminates the need to use expensive measures. It also has the potential to increase the effectiveness of healthcare services since it can take over some tasks and thus ease the workload for some healthcare providers. Telemedicine, on the other hand, will rise and become the order of the future since patient care will resemble a telecommunication network accessible to more patients, especially those in rural areas [34, 35]. It will also help in decentralized clinical trials such that more recruiting outlets can be used to enrol participants and obtain real-time data.

5.3 Predictive Wearables

The next generation of wearables will track health and forecast the onset of health problems. These devices could predict situations like heart attacks, allergic reactions, or any other complication which could be acted upon to prevent. One research area is enhancing the sensors to predict and monitor more aspects of human health, including blood sugar and oxygen saturation levels. Another focal is the mHealth application for diagnosing, preventing, and managing chronic diseases, including diabetes and hypertension [36]. Wearable technology can also deliver individualized advice on specific conditions and guide patients in making adequate lifestyle choices effectively. Another promising field is wearables for hospitals and healthcare facilities. With such devices, patient status can be observed continuously, with shifts in their condition being detected immediately and probably averting some deaths. Bright patches that help send body temperature readings and wristbands that help measure pulse rate and oxygen saturation will revolutionize in-patient care this year. Moreover, with devices that can be connected to our bodies and used for treating diseases, the opportunity of wearable technology in the healthcare sector includes the ability to use big data and AI and give everyone the unique, preventative medicine that they deserve. It is about patient-centred healthcare, which aims to help patients deal with their health and fitness independently to achieve better, faster, and more timely healthcare.

5.4 Smart Clothing and Implants

Bright clothing is one of the most prominent examples of emerging trends in wearable technologies. This technology integrates sensors and clothes to maintain records of the wearer's physiological factors and other related data. They include innovative jewellery with rings and necklaces that continuously monitor activities and heart rate. Clothing one wears to work out, sleep, and even lounge can be a smart health garment with sensors embedded in it. Such garments can monitor physiological activity such as heart rate, respiration rate, and body temperature and offer an overview of the health state when the individual conducts their routine work [37]. When discussing wearable tech, we often overlook one crucial category: implants. These are popularly referred to as 'The Invisible Health Protectors'. These devices, intended to be deployed in healthcare, are installed internally and offer sustained health checks, using no extra peripherals. Perhaps the most innovative of these is Elon Musk's Neuralink, a device that will be implanted in the brain to treat neurological diseases. Below is the possible use of this technology.

Parkinson's Disease Treatment: These implants inhibit affected brain cells and, in effect, could work to curtail the effects of Parkinson's disease.

Memory Enhancement: Such implants may also increase memory, useful in patients with Alzheimer's or any other type of dementia.

Mental Health Treatment: The technology may have a way of treating issues such as depression or anxiety since the circuits connect to the brain.

Another of the newest tendencies is the application of such technologies as augmented and virtual reality in medical center. These technologies can be used for analgesics, surgical rehearsals, and mental health disease management. The novelty of embedding health monitoring capabilities in smart clothing items will allow health monitoring without using health wearables. Also, new generations of implants, like neural interfaces for neurological disease treatment, will create new medical horizons.

6 Integration with Healthcare Systems

6.1 Wearable Tech for Chronic Disease Management

Wearable technology has been instrumental in managing chronic diseases. Examples include- Diabetes glucose monitoring, where the user and the healthcare givers can monitor glucose levels, and cardiac monitoring, using wearable ECG that lets the user and the doctors know when things are not well.

6.2 Personalized Healthcare: Tailored to the Patient's Needs

Another area in this regard is the healthcare function of wearable technology, which enables doctors to deliver personalized treatments to their patients. These medical devices can help provide specific and constant solutions for checking one's body condition. Culturally competent PHRs offer an optimal method whereby a healthcare professional comprehends a person's healthcare needs and can attend to them adequately and proficiently. Given the situation and growth of wearable devices, there is an excellent chance to create an app that can work simultaneously with these medical devices. It will help create a healthcare mobile application that effectively creates a healthy bond between technology and health.

6.3 Decentralizing Diagnostic and Therapeutic

It can be safely assumed that the diagnostics and therapeutics of the future will, to a greater extent, depend on wearable devices. As such, these devices can help in early diagnosis and individualized management of health conditions from outpatient services and hospitals.

7 Collaboration Between Tech and Healthcare Industries

Despite all the challenges witnessed today, consumer and medical-grade wearables will continue to increase in the coming years. That niche is still very profitable and has vast opportunities to contribute to the healthcare industry context. Stated apps will become more intelligent and devices more dependable, helping patients and clinicians to track an increased range of health symptoms. Thus, it is clear that even more industries and companies will begin the creation of connected devices and software solutions that will enable them. Hence, artificial intelligence in healthcare will continue progressing and influence wearables and their efficacy. It can improve the degree of accuracy of the measurements taken and decrease the threats from cyber assaults. Further, IoT technology and predictive analytical tools are best suited to be integrated to gain improved patient information [13]. New-age companies strive to stay on par with the increasing global demand and harness the benefits of medical wearables. Health care and wearable technology require tapping to deliver on the various needs one would offer.

7.1 Pros and Cons of Wearable Technology

Advantages of Wearable Technology in Healthcare

It is not a coincidence that many companies are using the development of technology as many opportunities and advantages have been seen in the relevant domain of the healthcare industry.

Promote Proactive Healthcare-Continuous Monitoring: Always On, Always Vigilant.

Another advantage of wearing devices in healthcare is that they enable constant tracking. These devices, which often have different types of sensors, can monitor many aspects of health at any time around the clock. This continuous monitoring of heart rates, blood pressure, and sleep patterns is a treasure trove of information to check for early signs of health problems. Health checks can be done daily; the moment signs appear; they are identified immediately. Consumers can take specific positive actions proactively to avoid severe health conditions at early stages. At the same time, a clinician or a doctor can assist the patient in rapidly reacting to the alerts and starting the treatment. Thus, wearables support improvements to the care process and take patient-centered approaches.

Boost Patient Engagement: Distributed technology allows patients to take personal responsibility for their well-being. Logically, having one's health records at a fingertip enhances the ability of an individual to make correct choices concerning lifestyle and treatment. Moreover, wearables are functional in enabling users to change for the better and adapt good behaviour throughout their lives [38]. With wearables, users get lots of valuable data, which allows them to monitor their health and get a sense of control [16]. The technology fosters a collaborative doctor-patient relationship and includes patients in managing their conditions successively. Although wearables are still in their infancy, healthcare providers can gain insight into utilizing the masses of health data collected by wearables. This real-time information enables patient-specific treatment, leading to elevated patient results. In addition, this data can be uploaded to the patient's EHR, giving the health professional the overall picture of the patient's health needs [29]. Thus, it aids in the utilization of data and enables one to make proper decisions.

Reduce Healthcare Costs: For the same reason, as wearables are directed towards better health, reduced disease vulnerability, and the means for early signs diagnosis, patients do not have to visit doctors and spend money on this. Healthcare providers can track patients after discharge to avoid accusing them of readmission charges or making losses by failing to reclaim from the insurers. In that way, the technology comes with positive returns of minimizing medical costs.

Decrease Staff Workload: Smartwatches and similar products create an enormous amount of personal health data that can be sent to physicians and used to track patients remotely [39]. It can assist in easing communication, minimizing face-to-face contact and balancing the working load of the members in the healthcare sector.

Role of AI and 5G: Artificial Intelligence and 5G technologies are improving the performance of Wearable Health care devices. Healthcare has been automated via AI with the help of machine learning algorithms, which diagnose and monitor patients' conditions closely and don't require interventions from doctors frequently. In medicine, for example, it can track deviant patterns in patients, which can call on data to predict and advise doctors in real-time. In the same breadth, 5G technology facilitates improved speed accompanied by low delay required for real-time remote patient monitoring and telemedicine [34]. It helps to enhance the health system and calls for early intervention in accident-prone healthcare situations.

7.2 Challenges of Implementing Medical Wearables

With the opportunities of wearable devices in healthcare come the issues that healthcare providers take note of today. The technology, albeit new, still has the benefits of improving. Four objectives must be highlighted here:

Technical Issues: Mobile health devices in the past have a high power demand, which is usually provided by long-lasting batteries [40]. The batteries should be compatible with wireless charging and integrated into compact devices to meet consumers' expectations. Hence, there is reason for companies to invest more in better semiconductor technology.

Data Accuracy: Wearable computers in healthcare must be as accurate and reliable as traditional computers. However, it can sometimes result in the wrong diagnosis or wrong treatment given to the patient, and in some circumstances, this may cause the death of the patient. Manufacturers, therefore, must ensure their devices are critically fine-tuned and endure tests that prove their accuracy [41]. Data is essential, but only when correct,this is the case with health data. That is why it must be created, transmitted, and worked through efficiently [42]. Here, the role of AI, Big Data, and IoT technologies can rise and support the process.

Security and Privacy: The owners of wearables are distraught about the safety and privacy of the health information they share. Despite this, it is still a barrier that should employ security protocols and observe the policy of the privacy of information when transferring and storing data [43]. HIPAA and the FDA recommended guidelines for Software as a Medical Device (SAMD) have made patient privacy no longer a choice but compulsory.

Improper Device Wearing: This is because the position of wearable sensors extensively affects measurements and data accuracy. They should be easily in direct contact with skin and positioned in the correct area. However, it is not always so with devices further secured with straps. Besides that, the results of measurements can be imprecise if moisture or even tattoos are on the skin. The proper integration of well-developed interfaces and operability is essential for wearables. These devices are easily compatible with Smartphone applications, cloud-based storage, and enhanced human interface. That is why maximally minimizing the friction of utilizing these devices can make using such devices a matter of course. Just think of having a wearable that can interact almost as a part of one's body. That is the kind of user experience that manufacturers strive to bank them. Nothing is more crucial to wearable health technology than ease of use, from synchronization to smartphone applications to data backup on cloud storage.

7.3 Overcoming Limitations and Challenges of Implementing Medical Wearables

Strategies to ensure data security

Data Encryption: The use of data encryption methodology while data is stored and while in transit forms an added security layer towards preventing data theft.

Regular Audits: It is suggested that audits be carried out regularly to detect any lapses that could have developed and take necessary action accordingly.

User Authentication: Sometimes, unauthorized personnel can access patient information, but strong user authentication measures will help eliminate such incidences.

Ensuring Accuracy and Reliability in Wearable Tech

Calibration: When used in calibration, it is made to give accurate results by successive calibrations of the devices used.

Quality Control: From the manufacturing procedures to the final output, quality is stringently maintained to obtain devices of excellent accuracy.

Clinical Validation: End products are typically clinically endorsed to ensure accuracy in an actual usage environment.

Accessible and Affordable

As wearable health technology becomes popular, the cost can still be prohibitive. Most high-end gadgets are so expensive they are barely affordable, especially to the average consumer. The key point for manufacturers of such devices is not to provide too many options and make devices cost considerably more than competitors, but at the same time to offer the basic set of options for their convenience.

Meeting Regulatory Compliance

All wearable health monitoring sensors are subject to compliance requirements beyond simple safety certifications and medical device approvals. Staying up with all of these rules could be involved, but it is paramount to promote the proper use of these gadgets.

Creating Seamless User Experience

Intuitive Design: Following with the interface and the device's design as one of the key elements of user experience. The interface has to be stylish and intuitive for the end-users.

Hardware Considerations: This is specifically a notable observation concerning wearable software. These applications should consider the hardware characteristics of the various platforms in question. For example, an Apple Watch app must account for the model and what sensors and options it has to provide the smoothest experience to the wearer.

Customer Support: Frequent and effective customer care services will boost the general usage or experience of the product or service being rendered to the users.

Tailoring Wearable Tech for Various User Scenarios

Concerning wearable software, it is confirmed that the common thing between them all is that they are not all similar. It is now the task of developers to imagine the numerous situations in which these devices can be utilized and designed for any number of front-ends. For example, while a typical wearables app appears as an icon on a smartphone, beyond the use of a more powerful device. A stable web application is required for problem-solvers who require vast system connectivity to health analytics [44]. It is all about building a software application that meets the needs of each end-user segment,whether worn around one's wrists, in a pocket or on a desk, the technology is as flexible as the health care it seeks to transform.

Tackling Battery Life and Comfort: The Unseen Challenges

Two significant problems in wearable technology are closely related to this issue: battery life and the comfort of wearing the device. There are already products with elements such as monitoring the respiratory rate—which needs better sensors and algorithms, which consumes power. Given the prospects and strict conditions described above, the development of wearable healthcare apps may be expected to be slightly above the average cost of typical health software.

8 Ethical and Regulatory Considerations for Equitable Access

A primary concern that arises in the creation of wearable technology for healthcare applications is that of data quality. Smart devices like blood pressure machines must consistently capture accurate data. The readings must be translated accurately to gain practical health recommendations. Wearable technology has social implications concerning the patient's right to privacy, protection of information, and self-determination [45]. Several issues related to the integration of wearable technology in health care, like legal frameworks for payment on wearable technology, patient data privacy, wearable data protection and adherence to standard specifications, need urgent attention [46]. Wearable technology devices capture and transmit personal health information, therefore, there is a need to ensure the privacy of such data to limited parties only.

Regulatory frameworks ensure these devices are safe, effective, and respectful of user privacy. Among the most influential are the U.S. Food and Drug Administration (FDA), the General Data Protection Regulation (GDPR) in the European Union, and the Health Insurance Portability and Accountability Act (HIPAA) in the United States.

8.1 FDA (U.S. Food and Drug Administration)

FDA defines wearable devices as medical devices and ensures their safety and efficacy. Several regulatory clearances or approvals exist for devices used in diagnosing, treating or preventing diseases, such as 510(k), De Novo classification and Pre-Market Approval 510(k)/PMA, etc. In addition, the FDA oversees mobile health apps and AI/ML algorithms when used in medical cases. Wearables such as CGM (continuous glucose monitors) are an area of focus for FDA, remote patient monitoring systems and smartwatches with ECG capabilities.

8.2 GDPR (General Data Protection Regulation)

GDPR is the central regulation for data protection law in the European Union. It also applies to any organization that processes the personal data of EU residents, regardless of the location of the headquarters in the EU province. It significantly impacts the development and deployment of wearables in Europe. Manufacturers need to safeguard data minimization, purpose limitation and privacy by design and balance transparency and user control with cybersecurity from the beginning. Since wearable devices can collect personal health data, it is considered "special category data" under GDPR, making it one of the highest levels of informed consent and protection.

8.3 HIPAA (Health Insurance Portability and Accountability Act)

The U.S.-based HIPAA secures individually identifiable health information in the hands of covered entities (e.g. hospitals/health plans) and their business associates. EHR synchronization is HIPAA-compliant, but HHS require Protected Health Information (PHI) with written rules in HIPAA for consumer-grade health devices. Wearables involving consumer electronics and clinical tools, resulting in requirements for unified access control. Coordinated with the collaboration of regulatory agencies like the FDA and GDPR enforcers, HIPAA overseers would create a slightly less disjointed structure that could benefit users. Additionally, areas like AI-driven diagnostics, cross-border data sharing, and the next wave of solutions in real-time remote monitoring require policy flexibility. For health tech to be practical and ethical, collaboration will be needed between technologists, healthcare practitioners, policy and legal experts.

9 Conclusion

Wearable health technology is rapidly evolving from simple fitness trackers to sophisticated, clinically relevant tools that are reshaping how healthcare is delivered and experienced. Thus, while still struggling and searching for the solutions, the tendency of using wearable tech in the sphere of healthcare is widely discussed, and it can be stated that this trend will bring about the essential transformation of patient care as precise, effective, and patient-oriented as far as possible. This chapter has traced the trajectory of wearable health tech, highlighting its growing role in continuous monitoring, early disease detection, and personalized care. The findings highlight the importance of key advancements such as biosensor integration, AI-powered data analysis, and seamless connectivity with healthcare systems demonstrate the immense potential of these devices to support preventive, remote, and patient-centered healthcare models. The implications are far-reaching like improved health outcomes through real-time interventions, reduced strain on healthcare systems, and increased patient empowerment. However, these benefits come with critical challenges, including data privacy, device standardization, and equitable access. As wearable technologies become more embedded in mainstream healthcare, clear policies and regulatory frameworks will be essential to ensure safety, efficacy, and inclusivity. There is also a pressing need for continued interdisciplinary research spanning healthcare, engineering, ethics, and policy to guide the responsible development and integration of wearables. The estimated expansion in five years based on the innovative advancement in AI, up-gradation of 5G technology will lead the incremental growth and projections of the size of the related market that will shift to a considerably more interconnected, preventive and individualized health system. Future efforts should focus on long-term clinical validation, addressing disparities in access and usability, and establishing robust health data governance. Ultimately, wearable health tech holds the promise to transform healthcare, but its full potential will only be realized through thoughtful innovation, collaboration, and policy development.

References

1. Smuck, M., Odonkor, C., Wilt, J., Schmidt, N., Swiernik, M.: The emerging clinical role of wearables: factors for successful implementation in healthcare. NPJ. Digit. Med. **4**(1), 45–48 (2021)
2. Coulter, A.: Engaging patients in health care. Open University Press, Maidenhead, Berkshire, United Kingdom (2011)
3. Werbrouck, A., Swinnen, E., Kerckhofs, E., Buyl, R., Beckwée, D., De Wit, L.: How to empower patients? A systematic review and meta-analysis. Transl. Behav. Med. **8**(5), 660–674 (2018)
4. Loncar-Turukalo, T., Zdravevski, E., Machado da Silva, J., Chouvarda, I., Trajkovik, V.: Literature on wearable technology for connected health: scoping review of research trends, advances, and barriers. J. Med. Internet. Res. **21**(9), e14017 (2019)
5. DeloitteInsights.: https://www2.deloitte.com/us/en/insights/industry/technology/technology-media-and-telecom-predictions/2022/wearable-technology-healthcare.html. Accessed 02 Dec 2024
6. Kim, J., Campbell, A.S., de Avila, B.E., et al.: Wearable biosensors for healthcare monitoring. Nat. Biotech. **37**, 389–406 (2019)
7. Golder, S.A., Macy, M.W.: Diurnal and seasonal mood vary with work, sleep, and day length across diverse cultures. Science **333**, 1878–1881 (2011)
8. Li, J., Xin, M., Ma, Z., et al.: Nanomaterials and their applications on bio-inspired wearable electronics. Nanotechnology **32** (2021)
9. Azodo, I., Williams, R., Sheikh, A., Cresswell, K.: Opportunities and challenges surrounding the use of data from wearable sensor devices in health care: qualitative interview study. J. Med. Internet Res. **22**(10), e19542 (2020)
10. Cheung, C.C., Krahn, A.D., Andrade, J.G.: The emerging role of wearable technologies in detection of arrhythmia. Can. J. Cardiol. **34**, 1083–1087 (2018)
11. Bayoumy, K., Gaber, M., Elshafeey, A., et al.: Smart wearable devices in cardiovascular care: where we are and how to move forward. Nat. Rev. Cardiol. **18**, 581–599 (2021)
12. Iqbal, S.M.A., Mahgoub, I., Du, E., et al.: Advances in healthcare wearable devices. Flex. Electron. **5**, 9 (2021)
13. Awotunde, J., Jimoh, R., AbdulRaheem, M., Oladipo, I., Folorunso, S., Ajamu, G.: IoT-based wearable body sensor network for COVID-19 pandemic. In: Advances in Data Science and Intelligent Data Communication Technologies for COVID-19. Springer, Cham (2022)

14. Seshadri, D.R., Li, R.T., Voos, J.E., et al.: Wearable sensors for monitoring the physiological and biochemical profile of the athlete. NPJ Digit Med **2**, 72 (2019)
15. Greiwe, J., Nyenhuis, S.M.: Wearable technology and how this can be implemented into clinical practice. Curr. Allergy Asthma Rep. **20**(8), 36 (2020)
16. Bove, L.: Increasing patient engagement through the use of wearable technology. J. Nurse Pract. **15**(8), 535–539 (2019)
17. Auerswald, T., Meyer, J., Von Holdt, K., Voelcker-Rehage, C.: Application of activity trackers among nursing home residents-a pilot and feasibility study on physical activity behavior, usage behavior, acceptance, usability and motivational impact. Int. J. Environ. Res. Public Health **17**(18), 1–21 (2020)
18. Radin, J.M., Wineinger, N.E., Topol, E.J., et al.: Harnessing wearable device data to improve state-level real-time surveillance of influenza-like illness in the USA: a population-based study. Lancet Digit. Health **2**, e85–e93 (2020)
19. Friel, C., Garber, C.: An examination of the relationship between motivation, physical activity, and wearable activity monitor use. J. Sport Exerc. Psychol. **09**, 1–8 (2020)
20. Dehghani, M., Kim, K., Dangelico, R.: Will smartwatches last? Factors contributing to intention to keep using smart wearable technology. Telematics Inform. **35**(2), 480–490 (2018)
21. Stiglbauer, B., Weber, S., Batinic, B.: Does your health really benefit from using a self-tracking device? Evidence from a longitudinal randomized control trial. Comput. Human Behav. **94**, 131–139 (2020)
22. Apple Watch.: Helping Your Patients Identify Early Warning Signs. Apple. https://www.apple.com/uk/healthcare/apple-watch/. Accessed 20 Nov 2024
23. Maeda, Y., Sekine, M., Tamura, T.: The advantages of wearable green reflected photo plethysmography. J. Med. Syst. **35**, 829–834 (2011)
24. Castaneda, D., Esparza, A., Ghamari, M., et al.: A review on wearable photo plethysmography sensors and their potential future applications in health care. Int. J. Biosens. Bioelectron. **4**, 195–202 (2018)
25. Maher, C., Ryan, J., Ambrosi, C., et al.: Users experiences of wearable activity trackers: a cross-sectional study. BMC Public Health **17**, 880 (2017)
26. Sivakumar, S., Bhatti, N.: Can smartwatch prevent sudden cardiac deaths? A case of smartwatch failure in arrhythmogenic right ventricular dysplasia. Cureus **13**, e15904 (2021)
27. Bent, B., Goldstein, B.A., Kibbe, W.A., et al.: Investigating sources of inaccuracy in wearable optical heart rate sensors. NPJ. Digit. Med. **3**, 18 (2020)
28. Molina-Garcia, P., Notbohm, H.L., Schumann, M., et al.: Validity of estimating the maximal oxygen consumption by consumer wearables: a systematic review with meta-analysis and expert statement of the INTERLIVE network. Sports Med. **52**, 1577–1597 (2022)
29. Ledger, D., McCaffrey, D.: Inside wearables: how the science of human behavior change offers the secret to long-term engagement. Endeavour Partners **200**, 1 (2014)
30. Hendker, A., Jetzke, M., Eils, E., et al.: The implication of wearables and the factors affecting their usage among recreationally active people. Int. J. Env. Res. Pub. Health **17**, 8532 (2020)
31. Keogh, A., Argent, R., Doherty, C., et al.: Breaking down the digital fortress: the unseen challenges in healthcare technology—lessons learned from 10 years of research. Sens. (Basel) **24**, 3780 (2024)
32. Liao, Y., Thompson, C., Peterson, S., et al.: The future of wearable technologies and remote monitoring in health care. In: American Society of Clinical Oncology educational book American Society of Clinical Oncology Annual Meeting, p. 115. NIH Public Access, Chicago (2019)
33. Chong, K.P.L., Guo, J.Z., Deng, X., et al.: Consumer perceptions of wearable technology devices: retrospective review and analysis. JMIR Mhealth Uhealth **8**, e17544 (2020)
34. Muniz-Pardos, B., Angeloudis, K., Guppy, F.M., et al.: Wearable and telemedicine innovations for Olympic events and elite sport. J. Sports Med. Phys. Fitness **61**, 1061–1072 (2021)
35. Hicks, L.L., Fleming, D.A., Desaulnier, A.: The application of remote monitoring to improve health outcomes to a rural area. Telemed. J. E Health **15**, 664–671 (2009)

36. Cho, I., Kaplanidou, K., Sato, S.: Gamified wearable fitness tracker for physical activity: a comprehensive literature review. Sustainability **13**, 7017 (2021)
37. Strain, T., Wijndaele, K., Dempsey, P.C., et al.: Wearable-device-measured physical activity and future health risk. Nat. Med. **26**, 1385–1391 (2020)
38. Neaţu, A.M.: Public health and behavioral economics. Nudging behaviors through wearable technology. Int. J. Econs. Prac. Theo. **5**, 518 (2015)
39. Seshadri, D.R., Drummond, C., Craker, J., et al.: Wearable devices for sports: new integrated technologies allow coaches, physicians, and trainers to better understand the physical demands of athletes in real time. IEEE Pulse **8**, 38–43 (2017)
40. Canali, S., Schiaffonati, V., Aliverti, A.: Challenges and recommendations for wearable devices in digital health: data quality, interoperability, health equity, fairness. PLOS Digit. Health **1**, e0000104 (2022)
41. Segura Anaya, L.H., Alsadoon, A., Costadopoulos, N., et al.: Ethical implications of user perceptions of wearable devices. Sci. Eng. Ethics **24**, 1–28 (2018)
42. Banerjee, S., Hemphill, T., Longstreet, P.: Wearable devices and healthcare: data sharing and privacy. Inf. Soc. **34**, 49–57 (2018)
43. Cilliers, L.: Wearable devices in healthcare: privacy and information security issues. Health Inf. Manag. **49**, 150–156 (2020)
44. Martin, S., Kelly, G., Kernohan, W.G., McCreight, B., Nugent, C.: Smart home technologies for health and social care support. Cochrane Database Syst. Rev. (2008)
45. Tu, J., Gao, W.: Ethical considerations of wearable technologies in human research. Adv. Health Mater. **10**, e2100127 (2021)
46. Nebeker, C., Harlow, J., Espinoza Giacinto, R., et al.: Ethical and regulatory challenges of research using pervasive sensing and other emerging technologies: IRB perspectives. AJOB Empir. Bioeth. **8**, 266–276 (2017)

Monirul Islam With over 1 years of combined industrial and academic experience, Dr. Monirul Islam is an Associate Professor in the Department of Management at Sister Nivedita University, Kolkata. Specializing in marketing, his research interests include AI/ML, climate change, SDGs, healthcare innovation, and psychological stress. Dr. Islam has published over 30 journal articles and contributed 15 book chapters to prestigious publishers including Scopus, ABDC, Emerald, Elsevier, Springer, Springer Nature, Wiley, and Taylor & Francis. He has served on editorial boards and review committees for SCOPUS and UGC-CARE indexed journals and contributed to several AICTE-funded projects.

The Future of Wearable Devices in Healthcare

Vikrant Verma and Charu Saxena

Abstract Wearable devices have transformed healthcare by allowing constant monitoring of health, early diagnosis, and personalized treatment. In this chapter, the transforming role of wearable technology in clinical decision-making, preventive health, and patient care is discussed while evaluating its impact on remote patient monitoring and chronic disease management. It analyzes the intersection of artificial intelligence (AI), predictive analytics, and sensor technology in healthcare applications, regulatory barriers, data privacy issues, and emerging trends in wearable healthcare technology. Sensors, AI, and connectivity technologies are revolutionizing wearables from static vital sign monitors to predictive health devices that can detect potential health problems in real time. Devices like biosensors, fitness bands, and smartwatches are increasingly being used to care for chronic disease, allowing patients and clinicians alike to use data to remotely adjust treatments. The technologies hold the potential to enhance outcome by better engaging patients and improving compliance with professional advice. The chapter also tackles key challenges such as data privacy, regulatory frameworks, and the necessity for robust validation of wearable health metrics and delves into inequalities in access to wearable technology and options to address them. Through an examination of emerging trends, implementation issues, and future trajectories, this chapter presents a broad view of how wearable technology will influence healthcare systems, developing responsive and individualized health management systems that hold the promise to profoundly improve global public health outcomes.

Keywords Wearable technology · Healthcare innovation · Remote patient monitoring · Personalized medicine · Health data privacy

V. Verma (✉)
Faculty of Pharmacy, Department of Pharmaceutical Chemistry, Swami Vivekanand Subharti University Meerut, Meerut, India
e-mail: pharmacy@subharti.org

C. Saxena
Faculty of Pharmacy, Department of Pharmaceutics, Swami Vivekanand Subharti University Meerut, Meerut, India

P. Eappen et al. (eds.), *Remote Monitoring and Wearable Devices in Healthcare*, Information Systems Engineering and Management 63,
https://doi.org/10.1007/978-3-031-98897-4_11

Wearable Technology: A Brief History and Types

Wearable technology has developed significantly after many years since its emergence. In the early days, devices like pedometers and elementary heart rate monitors were restricted to tracking physical activity and vital signs. But with new breakthroughs, advanced sensors can now track blood oxygen levels, glucose levels and even stress levels. The main focus of these movements is on enhancement of sensor technology, integration of AI and superior connectivity alternatives.

Wearable devices have revolutionized the healthcare landscape at an unprecedented pace by facilitating real-time health monitoring, early diagnosis and tailored treatments. The change is primarily due to technological development, the incorporation of AI (Artificial Intelligence) and machine learning for the functioning of these devices [1, 2].

Wearable Devices: Definition and Types

Wearable technology is a broad term that refers to any device that you wear to track certain health metrics. These include:

Smartwatches: Which have sensors to monitor heart rate, physical activity, and sleep metrics.

Fitness Bands: These are designed to monitor physical activity and calories burned.

Biosensors: Analytical tools that translate biological reactions into electrical signals so that particular biological, chemical, or physiological parameters, like blood pressure or glucose levels, can be detected and tracked.

Smart Clothing: Clothes that have sensors embedded in them to monitor physiological parameters such as heart rate and respiratory rate.

These devices can gather data in real-time as well as on pre-specified intervals, which allows for a holistic view of the patient's health state [3, 4].

1 Role of Wearable Technology in the Future of Health Care

1.1 Continuous Health Monitoring

Wearable technology enables real-time health tracking, enabling constant access to crucial health indicators like heart rate, blood pressure, and physical activity for the patient and the health care provider. The continuous accumulation of this data is necessary in order to detect possible health problems early and take appropriate actions in good time. For example, smartwatches and fitness trackers can tell users if their heart rhythm is irregular, or if there has been a substantial change in their overall activity levels, so users can consult a physician before they get sick [5].

Wearables also offer capabilities for continuous monitoring, which can be particularly useful in treating chronic diseases. Wearables monitor diabetic or hypertensive patients and provide real-time feedback about their current state of health, allowing them to modify their treatment plan proactively, such a transformation from reactive to proactive health care is a significant breakthrough in modes of patient management [6].

1.2 Predictive Analytics and Early Diagnosis

Smart wearable devices not only serve as monitoring devices but also utilize predictive analytics based on AI which is the application of machine learning methods and statistical algorithms to determine the probability of future events based on patterns in historical data. A 2023 WHO report titled Digital Health Interventions highlighted that wearable devices that can identify abnormal heart rate variability (HRV) (The difference in heartbeat duration, a measure of cardiovascular health) through AI and decrease the risk of sudden cardiac events by 30% with early intervention. These technologies identify trends over time, indicating health problems in their initial stages. Additionally, research in Nature Digital Medicine done during 2022 proved that wearable ECG (electrocardiogram) monitors with AI capabilities and record the heart's electrical signals and identify anomalies with 92% predictive accuracy for atrial fibrillation (AFib) events, which are fast and jumbled electrical signals in the heart's upper chambers that indicate an irregular heartbeat, before they take place. Predictive software, such as those reviewing HRV patterns, are breaking new grounds in detecting early cardiovascular disease.

A major development in patient management techniques is the transition from reactive to proactive healthcare [7]. Integrating wearables with electronic health records (EHRs) also improves the availability of patient data for providers. Such data integration also supports early diagnosis and personalized treatment plans, enabling clinicians to make informed decisions based on comprehensive and up-to-date information through seamless data flow [8].

1.3 Personalized Treatment

Wearable technology offers one of the transformative shifts based on personalized health care through customization in which recommendations are personalized. AI-based wearables adapt them according to the user data. This could include personalized recommendations, such as suggestions for exercises or dietary changes based on a user's specific health data or goals. Patients are thus able to take an active part in the management of their health by exploiting this degree of personalization, which results in better adherence to treatment plans and better outcomes.

Wearable technology also plays a significant role in telehealth initiatives by enabling remote consultations, whereby doctors can monitor a patient's condition in real-time without the patient needing to visit a clinic. This helps people who live far away or people with limited mobility [9].

1.4 Remote Patient Monitoring in Patient-Centered Care

Remote patient monitoring (RPM) using wearable technology has become a revolutionary method of healthcare delivery, especially in managing chronic diseases. The method allows healthcare professionals to keep patients vital signs and health measures under constant observation without the need for constant clinic visits, which makes it most beneficial for patients with mobility problems or those in remote locations. Clinical evidence on the effectiveness of RPM is increasing. In 2022, a Mayo Clinic pilot study provided chronically heart failure patients with AI-paired ECG-monitoring smartwatches that lowered emergency department visits by 30% [10]. This is empirical proof of how hospitalizations can be lessened and patient outcomes enhanced by wearable-based RPM. Likewise, a systematic review of 42 randomized controlled trials reported that RPM decreased hospital admissions by 23% in patients with chronic respiratory disease and 29% in those with cardiovascular disease [11].

Clinical workflows wearable devices give healthcare practitioners actionable information, enhancing decision speed and accuracy. Wearables data integration with electronic health records presents a holistic overview of patient health status, making it possible for more tailored treatment plans and treatment options. The use of AI-enabled wearables in hospital workflows enhanced early detection of diseases by 20%, as indicated by a study that was published in the Journal of Medical Internet Research [12].

Beyond clinical outcomes, wearable-based RPM enhances the patient experience through:

Convenience: Patients can manage conditions from home while maintaining provider contact.

Empowerment: Real-time access to health data encourages active participation in care.

Personalization: Continuous monitoring enables treatment adjustments based on individual responses.

Safety: Early detection of deterioration allows for timely interventions.

The shift to home care enabled by wearable technology also has important economic advantages. A cost-effectiveness analysis determined that implementation of RPM lowered total healthcare costs by USD 8375 per patient per year for patients with congestive heart failure and USD 5430 for patients with chronic obstructive pulmonary disease.

As RPM technologies advance, their integration with telehealth services provides a holistic virtual care system that preserves the human touch between patients and providers while taking advantage of the accuracy of continuous data collection [13].

2 Top and Current Trends of Wearable Technology

As wearable technology advances, some new trends are defining its future in healthcare. With advancements in textile technology, smart clothing has created garments with sensors able to monitor vital signs and additional health data. They offer a non-intrusive means for users to monitor their health, with no extra hardware required.

Augmented and Virtual Reality These technologies are being tested for a range of cases in healthcare, including pain management and surgical training. Immersive environments created with wearable AR/ VR devices aids in enhancing the patient experiences and improves the outcomes of training for the healthcare professionals.

Integration with IoT The Internet of Things (IoT), a system of physically linked devices that are connected and equipped with sensors, software, and connectivity to gather and share data, allows wearables to interact with other devices and systems in the healthcare landscape. This connectivity enables real-time data exchange between patients and providers, enabling prompt interventions and improved care coordination [14].

The future of wearables will have personalized health recommendations. Such suggestions might include lifestyle changes or warnings for health risks unique to each user's profile [15].

Advanced Health Monitoring.

The trend of health monitoring capability is one of the trendiest technology trends in wearable technology. Modern wearables are powered with smart sensors that can constantly monitor certain vital signs specifically heart rate, blood pressure, oxygen saturation and glucose levels. For example, the Apple Watch Series 9 is expected to have non-invasive glucose monitoring, which will allow people with diabetes to manage their condition without having to prick their fingers with a needle every few hours. It enables users to make proactive decisions about their health, providing health professionals essential information for diagnosis and care [16].

Growth Opportunities in Healthcare Applications.

Wearable technology in healthcare is growing rapidly in fact, wearables are poised to be a cornerstone for remote patient monitoring (RPM), as healthcare providers monitor their patients' vital signs and other health metrics in real time. This trend is particularly good for chronic conditions like heart disease and diabetes since tracking them 24/7 can be crucial to proper management. The Worldwide Wearable Market for Health & Wellness is expected to experience a market growth [17].

Interoperability and Improved Connectivity.

There is a growing emphasis on improving connectivity and interoperability between different platforms and devices which is the capacity of various devices, applications, and information systems to access, exchange, integrate, and use data cooperatively and synchronizedly as wearable technology becomes more widely used.

By connecting securely to a smartphone, health apps and electronic medical records (EMRs), wearables do not require manual interaction but can share data towards the patient overall status. This interoperability is important for coordinating healthcare providers to give care along with ensuring that patients get the treatments they need as a result of their health data [18].

3 Sensors and Artificial Intelligence (AI) and Connectivity Innovations

3.1 Innovations in Wearable Sensor Technology

Current wearables employ a range of sensors to measure physical, chemical, and biological signals. Recent clinical validation research has shown substantial progress in sensor technology. For instance, a detailed validation study of wearable piezoelectric blood pressure sensors (WPBPS) showed excellent performance characteristics such as high normalized sensitivity (0.062 kPa^{-1}) and rapid response time (23 ms) for continuous non-invasive arterial pressure monitoring. This 35-subject clinical trial in the age range of 20–80 years validated that these sensors are capable of attaining accuracy on par with conventional medical-grade devices, with mean differences of $\pm$ 5 mmHg for systolic and $\pm$ 3 mmHg for diastolic readings relative to standard cuff-based measurements. Empirical validation of this nature is essential for wearable technology to move from consumer devices to trustworthy medical devices [19].

AI-enabled wearable sensors are revolutionizing health monitoring by studying user behavior patterns and offering predictive health information. For instance, AI-powered continuous glucose monitors (CGMs), a system that continuously measures the amount of glucose in the interstitial fluid day and night and provides information on trends and patterns in the amount of glucose not only detect blood sugar levels but also anticipate hypoglycemic events 60 min ahead of time, enabling diabetic patients to act in advance. A 2023 study published in The Lancet Digital Health revealed that AI-enabled CGMs enhanced hypoglycemia detection accuracy by 40%, drastically minimizing diabetes complications.

1. These innovations contribute to the continuous monitoring of health metrics and allow healthcare providers to monitor patients conditions more effectively.

Types of Sensors:

Physical Sensors: Verify conditions like heart pace, temperature, and movement.

Chemical sensors: chemical biosensor analyzes metabolic health biochemical markers.

Biosensors: Observe physiological changes that pertain to particular health conditions [20].

3.2 The Role of Artificial Intelligence

Combining AI into wearable technology represents a major stride in the field of healthcare. Wearables generate huge quantities of data which can be processed in real time by AI algorithms. This enables real-time analysis and interpretation of health parameters, thus allowing timely intervention once an abnormality is identified. AI can also look at trends in heart rate variability to predict possible cardiovascular events before they happen, for example [21, 22].

3.3 Connectivity Enhancements

Wearable devices did not just stop with the fitness; their connection functionalities have also been constantly emerging. Most of the modern wearables use Bluetooth or Wi-Fi to transfer data to smartphones or cloud-based platforms. Such connectivity also enables seamless integration with electronic health records (EHRs) allowing healthcare providers to access in-depth patient data and history remotely. This allows clinicians to make timely decisions based on up-to-date information on the health status of their patients [23].

4 Wearable Devices in Chronic Disease Management

Wearables including smart watches, fitness bands, biosensors are becoming the ultimate tool for monitoring chronic diseases. Thus, these technologies help patients and healthcare providers remotely make data-driven changes to treatment plans and improve overall care quality.

Chronic Health diseases like diabetes mellitus, Hypertension, CVD and Respiratory disorders brings a lot of burden to the healthcare systems. These conditions are often managed through continuous monitoring and timely interventions to avoid complications. When it comes to these challenges, wearable devices have appeared as the demanded solution offering the real-time health-related data that can be readied as valuable data while making treatment decisions [24].

4.1 Improving Management of Chronic Conditions

Wearables empower patient by advocating self-management of their chronic conditions. Hence, these devices offer actionable information about the health status of the users which motivates them to be active in managing their health status. For example, wearables can gently prompt patients to take their medication or let them know when it's time to move. More engagement leads to increased compliance to treatment protocols and health outcomes [25].

4.2 Applications in Chronic Disease

Wearable technology has been clinically tested and used successfully in a broad array of chronic diseases with empirically proven effects. One validation study tested the reliability of wearable devices for health data interpretation in many chronic conditions. The researchers gathered more than 15,000 data points from 87 individuals with many chronic conditions such as diabetes, hypertension, and cardiovascular disease. In comparison with clinical gold standards, the calibrated wearables produced 93% accuracy for heart rate measurement, 89% for sleep quality rating, and 91% for activity categorization. What was most important to observe from the study, however, is that device accuracy significantly differed according to context of use, and this accuracy was greater for sedentary behaviors (95%) than for strenuous movement (82%). The investigators set statistical criteria for clinical acceptability: correlation coefficients > 0.85 with reference standards, mean absolute percentage errors < 10%, and intraclass correlation coefficients > 0.75 for test–retest reliability. These empirically established quality standards give healthcare professionals clear metrics to assess the clinical utility of particular wearable devices [26, 27].

Diabetes Management

Wearable Devices Continuous monitoring of blood sugar level is a core facet of diabetes management. At the same time, CGMs are changing the way patients manage their disease by providing real-time data to avoid hypoglycemia and hyperglycemia. Research has demonstrated the benefit of CGMs with better glycemic control with lower A1C in individuals with type 1 and 2 diabetes.

Cardiovascular Health

Wearables are used in cardiology for monitoring heart rate variability, arrhythmia detection, and hypertension management. Devices with ECG can detect irregular heart rhythms (known as arrhythmias) that should be evaluated by a healthcare professional. Research has shown that wearables can greatly improve cardiac rehabilitation by offering patients feedback about their exercise regimens and general heart health [28].

Respiratory Condition

In patients with respiratory conditions like asthma or chronic obstructive pulmonary disease (COPD), wearables can track lung function and identify exacerbations early on. Home spirometry and pulse oximeters provide valuable data which enables patients to manage their symptoms better. Wearables have been shown to improve adherence to prescribed medication regimens and decrease hospitalizations for respiratory illnesses.

Mental Health Monitoring

Wearable technology is also being explored as an option to help with mental health management. Open endocrinology devices with sensors to monitor physiological responses (e.g., heart variability) can be used for monitoring stress levels or anxiety episodes in real-time. That means people can put coping strategies into place, or go for professional help when they need it [29].

4.3 Advantages of Using Wearable Devices for Management of Chronic Diseases

Using wearable technology in chronic disease management has several advantages:

Improved Patient Engagement

Wearable devices drive deeper patient engagement by offering users instant access to their health information. Such transparency allows individuals to take ownership of their health and adhere to their treatment plans so compliance with treatment plans accurately measures adherence to treatment plans. Studies have shown that greater engagement improves health for many chronic conditions.

Cost Reduction

Wearable devices can also potentially decrease healthcare costs resulting from emergency visits and hospitalizations by allowing for remote monitoring and timely interventions. Patients who take an active role in managing their conditions with the help of wearables, are less likely to develop complications that need costly medical procedures. Health management practices with wearable devices could reduce healthcare costs for people with cardiovascular diseases as much as 12% depending on their type of heart disease and presence of related conditions discovered in a study published by the American Heart Association [30].

Enhanced Data Collection

Wearables enable the collection of immense amounts of health data over long periods, which in turn affords additional opportunities to correlate comprehensive patient outcomes. This amount of information can be vital for clinical research and optimizing treatment protocols. It enables better trend analysis in patient data to improve clinical decision-making process [31].

5 Future of Wearable Technology: Challenges in Implementation

While there are many advantages of the use of wearable technology in chronic disease management, several issues are still the need of the hour:

Data Privacy and Security

Collecting that kind of sensitive health information raises questions of data privacy and security. Protecting patient information from unauthorized access is of utmost importance in sustaining the trust patients have placed in the healthcare system. As technology progresses, the regulatory frameworks have to adapt accordingly to ensure that the patients information is protected.

Accuracy and Reliability

Wearable data is prone to numerous noise factors and erroneous readings, necessitating stringent validation. A systematic review of 24-h physical behavior measurement using wearables under laboratory conditions found key methodological considerations for validation studies. The review of 83 validation studies revealed that only 27% of commercially available devices had acceptable accuracy when compared to gold-standard laboratory measures. The authors determined that validation protocols should include: (1) comparison with criterion measures in controlled environments, (2) testing in diverse populations including those with compromised movement patterns, (3) testing under free-living conditions, and (4) clear reporting of data processing algorithms. The review concluded that numerous devices now being marketed for healthcare use are not adequately empirically validated, pointing to the nonalignment between commercial claims and scientific evidence. For clinical use, devices must show less than 5% error rate in controlled environments and retain accuracy within 10% [32–34].

Incorporation into the Healthcare System

The incorporation of this type of data into established healthcare systems has logistical challenges. The wider access to data will force those managing patient care to create workflows where this new source of information actually gets used, while ensuring that devices can interoperate. By 2023, the wearables market is expected to reach $60 billion, but effective integration is needed to realize the full potential of wearables in clinical practice [35].

5.1 The Future of Wearable Technology in Chronic Disease Management

As technology continues to evolve, here are a few trends that are expected to drive the future of wearable devices in healthcare:

Increased Interoperability

Aspects to improved interoperability across various wearables will enable seamless data transfer across devices. Healthcare systems will then be able to access full overviews of patient health across monitoring networks.

Artificial Intelligence Integration for Predictive Insights

Artificial Intelligence (AI) will add predictive analytics capabilities in wearable technology. From wearables, vast data is collected, and AI algorithms can be used to analyze the data and accordingly help identify patterns that indicate a potential healthcare issue before it becomes critical. This will in turn add to the empowerment of both patients and providers.

Adding New Therapeutic Areas

As research continues into various therapeutic areas for which wearable technology could be applicable, we might witness proliferation into therapeutic areas as diverse as oncology or neurology. Wearables may be useful in tracking treatment responses or warning of early complications from cancer therapies or of neurological disorders [36].

6 Growing Recognition of Wearable Technology

The growing recognition of wearable technology such as smartwatches, fitness and activity trackers, and biosensors as avenues to improve patient engagement and compliance with medical recommendations is realized. These devices give patients the ability to monitor their health in real-time as well as provide incentive towards being proactive in regards to health management which ultimately puts patients in charge of their health outcome [37].

6.1 Enhancing Patient Engagement

Real-Time Data Access

A leading benefit of wearable devices is the ability for patients to access their health data in real-time. This transparency enables individuals to take an active role in their care and shapes a sense of ownership over one's health. Wearables can track vital signs like heart rate, activity levels, and sleep patterns, enabling patients to monitor their metrics over time. Studies show that when patients can see their health data, they make more informed decisions about lifestyle changes and treatment decisions.

Personalized Health Insights

Wearable devices generate a wealth of data that can be interpreted into individualized health information. By seeing each individual's health habits, these healthcare providers are able to create personalized interventions and advice. A good example could be that a smartwatch may inform a user about an increased heart rate or inactivity which forces the user to take corrective measures. This personalized feedback not only promotes patient engagement but also strengthens ownership of one's health.

Increased Accountability

Patients are motivated to partake in programming and are held more accountable to themselves because they tend to see themselves moving toward their health goals. Some features like setting goals and tracking achievements encourages users to stay to their prescribed medication and healthy activities. Research has shown that if patients are able to track their progress whether that's hitting daily step goals or keeping glucose levels stable they are more likely to stay motivated to meet their health targets. Preventing the number of complications or exacerbations of chronic disease is very much a matter of compliance with the treatment regimen, and even more so with chronic diseases, so this kind of accountability can have a very high impact [38].

6.2 Promoting Compliance with Medical Guidance

Ongoing Monitoring and Notifications

Wearable technologies allow for constant monitoring of health metrics, which can improve compliance with medical recommendations. For example, biosensor-equipped devices can recognize abnormalities in physiological properties and promptly notify users. One application of this is enabling patients to act proactively to address potential health issues, such as obtaining care during an episode of atrial fibrillation, which in turn takes place by enhancing compliance to prescribed medical therapies [39].

Integration with Healthcare Providers

Wearables facilitate seamless patient–provider contact by streaming data directly to a clinician in real time. By continuously collecting data on the patient, healthcare professionals are able to remotely track high-risk patients, and react appropriately, changing their treatment plan when the need arises, based on the latest available data. This integration contributes not only to patient safety but also to a collaborative partnership between patients and providers [40].

Education and Empowerment

Wearable technology also has an educational component that is often ignored. Wearables enable users to understand how their daily behavior affects their health, thus giving them greater autonomy in choosing the right lifestyle and treatment when available. Wearable apps with educational features can also help people learning the importance of sticking to medications schedule or a recommended diet which further helps in encouraging quality behaviors [41].

6.3 Implications for Health Care Delivery

Wearable technology in healthcare has a long-term impact on the patient engagement and adherence. In this Patients are the Center of Focus and principle is shifted towards Patient-Centered Care. This capability is pivotal to a more patient-centric evolution of healthcare allowed by wearable devices. Wearables play an integral role by putting health data at patient's fingertips and encouraging them to take an active role in their health management. This transition not only increases patient satisfaction but improves clinical outcomes as patients become more involved in their care [42].

7 Data Privacy, Breaches, Regulatory Aspects, and Validation

7.1 Data Privacy Concerns, Risks of Data Breaches and Legal Environments for Information Assurance

The Nature of Data Collected

Wearable devices can gather a wealth of sensitive health information, including heart rate, physical activity, sleep patterns, also biochemical data through biosensing. This data is sent wirelessly as a stream to cloud platforms, where it can be analyzed and stored. This data is sensitive, and its unauthorized access or data breaches can result in identity theft or the misuse of personal health data which raises major privacy concerns.

The data breaches of wearable devices can lead to dire consequences. This access could enable hackers to use a user's health data for malicious means, including targeted advertising or discrimination by an insurance company. According to a study, 80% of customers are worried about how third parties gather and use their personal health information. The broad use of wearable technology in healthcare may be hampered by this mistrust [43, 44].

To mitigate these concerns, regulatory frameworks must be put into place. Existing laws like the Health Insurance Portability and Accountability Act (HIPAA) in the US a US federal law that established national guidelines to prevent the disclosure of private patient health information without the patient's knowledge or consent, establish baseline protections, but as wearables continue to revolutionize healthcare, they fail to fully address new complications that arise. Policymakers need to create robust rules that uniquely apply to the challenges surrounding wearables such as data ownership, consent management, and third-party access [45].

Regulatory Issues

Lack of Standardization

This could have perhaps been one of the biggest regulatory roadblocks for wearable technology. Most of the wearable devices are marketed as consumer products, not medical devices, there are variability in quality and accuracy of data. Without the standardization of protocols for validation and testing, these devices could lack the necessary safety and efficacy requirements.

Navigating Regulatory Landscapes

Wearable devices need to comply with regulatory norms before they can be accepted widely in clinical use. For instance, the FDA's Software as a Medical Device (SaMD) framework regulates AI-based wearables, validating safety and efficacy before they are made available to consumers. The European CE marking process under Medical Device Regulation (MDR) also requires rigid compliance for remote monitoring applications.

One of the major challenges facing the industry is how to traverse these regulatory environments to attain global acceptance. Whereas GDPR which is alegal framework that establishes rules for gathering and using personal data belonging to EU (European Union) citizens, protecting data in Europe, differences in international standards present obstacles for manufacturers of devices. Cooperation between regulatory agencies, healthcare professionals, and technology companies is needed to harmonize these systems and expedite the uptake of safe, effective wearable technology [46–48].

7.2 Stakeholder Collaboration

Collaborative efforts among stakeholders (manufacturers, healthcare providers, regulators, and patients) are critical to overcoming regulatory challenges. Defining specifications for device creation and data handling has portrayed solutions in terms of making wearables safe as well as protecting private information. Partnerships also enable the development of industry standards that can accelerate interoperability between devices and healthcare systems [49].

7.3 *The Importance of Rigorous Validation of Wearable Health Metrics*

The Need for Accuracy and Reliability

Data collected from wearables play an important role in clinical decisions and, therefore, their accuracy and reliability are of utmost importance. False readings can result in inappropriate treatment decisions or unnecessary anxiety for patients. For example, in one study, inaccurate measurement in heart rate monitoring may be due to motion artifacts or in the wrong position of the device. It is critical for user and health care provider trust that wearable devices provide accurate measurements.

Uniformity of Validation Processes

A standard for validating wearable health metrics performance is being developed through recent empirical studies, although worldwide agreement continues to be in the making. A clinical validation study suggested an all-encompassing framework for assessing wearable devices that has found much support from researchers and regulatory agencies. The study cross-validated an ear-worn gait analysis device against laboratory-based motion capture systems across 42 participants with varied mobility profiles. The validation procedure developed a three-tier assessment framework: (1) technical validation of raw sensor outputs against calibrated instruments, (2) analytical validation of derived measures against gold standards, and (3) clinical validation investigating real-world utility in medical decision-making. The researchers established that whereas the device met acceptable technical accuracy (within 2.3% of reference standards), clinical validity was very much context-dependent upon the particular gait parameter and user population. For example, the accuracy for stride length measurement was up to 97% in healthy adults but fell to 82% among individuals with mobility impairments. The empirically-derived validation framework gives manufacturers and researchers clear benchmarks for demonstrating device reliability across different contexts of use [50]. It is essential to have rigorous validation processes in place to set benchmarks for accuracy and reliability. In the meantime, researchers call for standardized protocols to enable uniform testing and assessment of wearable technologies before market entry [51].

Clinical Trials and Research Studies

Wearables assigned to patients could also generate meaningful clinical data that helped describe their efficacy in the practical environment. But a significant number of studies are based on consumer-grade wearables, which may not meet medical-grade standards. Research using validated medical devices will produce more robust outcomes data.

Moreover, continued advancement in sensor technology and calibration algorithms will further improve the accuracy of wearable devices. Financing for research will tackle existing limitations while inspiring innovation in this fast-moving realm [52, 53].

8 Disparities in Access to Wearable Technology

Wearable technology linking with health care may prove remarkable in revolutionizing in management of disease and treatment of patient Nevertheless, disparities in access to those devices represent great challenges that should be overcome to guarantee health equity for all population groups. The current gaps and inequities in the access to wearable technology, and the related factors leading to those inequities, in addition to strategies to improve accessibility and promote health equity, are discussed here [54, 55].

Socioeconomic Factors

Wearable technology is significantly distributed based on socioeconomic status. People from high-income groups are more likely to have the financial means to buy higher-end wearable devices, whereas individuals belonging to lower-income groups may not even possess the basic models. Recent research indicates that the price of wearable technology continues to be a major deterrent to adoption, with many people showing interest in fitness trackers but being unable to buy them because of budgetary limitations. This create a situation in which only populations with sufficient economic resources are able to enjoy the benefits brought by the advancement of health monitoring technologies, thereby creating a vicious cycle by which the gap of health disparities are widening between people getting rich because of these technologies and those exposed to socio-economic disadvantages [56].

Educational Barriers

Educational barriers affect access to wearable technology too, in addition to financial constraints. For example, many people do not know how to use wearables or interpret the data they provide. According to a survey, more than 51% of adults with health insurance struggle to comprehend their coverage, which also includes being unclear about the advantages of the various health technologies that are available. In the absence of strong educational efforts to teach patients and future patients how to use such devices and what they mean for their health care, many people who could potentially benefit from wearable technology will likely miss out on it [57].

Geographic Disparities

Tracking-wearable access disparities are also compounded by geographic location. Remote and underserved urban environments frequently exhibit inadequate access to technological resources, particularly in healthcare. The other major barrier in these regions is the availability of internet connectivity, which is also a requirement for most wearable devices. We thus need to ensure a balance between transformative technologies as tools for preventative health while also being mindful of growing disparities in health and healthcare access due to the digital divide, which leaves vulnerable populations without the very tools to facilitate effective health monitoring and management across the health continuum [58].

8.1 Strategies for Increasing Accessibility

Here are some ways to solve these differences and improve access to wearable technology:

Cost-Effective Manufacturing

Mass production of wearables is one of the most straightforward methods of increasing wearables accessibility. Manufacturers can invest in more basic, less-feature models that are more reasonably priced but provide many of the basic functions. With the potential to drive down production costs by offering lower-priced alternatives, companies can market wearables to a wider range of consumers. What's more, partnerships with local healthcare organizations can help with distribution in underserved areas, ensuring the people who need these technologies the most can access them [59].

Subsidies and Financial Assistance Programs

Subsidizing or providing financial assistance for these programs is another measure that can reduce cost related barriers. By partnering with manufacturers, governments and healthcare organizations can create programs that offer incentives to reimburse the cost of wearables for people with low income. These programs could include wearables as part of preventive health programs or chronic disease management initiatives. By reducing the cost barrier associated with the purchase of such devices, there will be a wider base of users benefitting from their application [60].

Patient Education Programs

Patient education should be reinforced for higher accessibility and engagement of wearable technology. Educational programs may be needed to explain the utility of wearables, how they work, and how they can be easily incorporated into daily health management strategies. Workshops, online courses, and informational materials can go a long way in bridging this knowledge gap, empowering citizens to take full advantage of existing technologies.

Community-Based Initiatives

Building community-based systems to increase access to wearable technology for underserved populations. Wearables can be distributed from local health centers and community organizations along with education on how to use them. Partnerships with pharmacies or community health workers can help with outreach to residents to educate them on the technologies available and help them access devices [61].

Research Inclusion

It is imperative that research studies applying wearables include individuals from all socioeconomic strata to assure the technology can support population based needs. Clinical trials have historically excluded underrepresented groups from participation, resulting in products that might not have met their needs fairly enough. By integrating inclusivity from the start when designing research studies, manufacturers will be able to develop wearables which are more pertinent and successful for everyone [62].

What Policy Can Do to Advance Equity

Regulatory Changes

Regulatory changes are needed to enable fair access to wearable technology. Designing products to be accessible can be driven by regulation (e.g. policies and standards) that rewards manufacturers who make efforts to create accessible products. Guidelines could also be established that mandate diversity in populations included in clinical trials, ensuring that the data collected reflects a wider range of experiences and needs [63].

Insurance Coverage Expansion

Covering wearable devices through insurance is another important avenue to improving accessibility. The cost of buying wearables or using them as part of a treatment plan isn't covered by many insurance plans today. This reduction of financial barriers to wearable technology by policy-change is not without some effort, as significant advocacy may need to occur before wearables are considered to be legitimate medical tools that are worthy of reimbursement [64].

9 Future for Wearable Technology

As wearable technology advances, we expect several trends to emerge:

Healthcare Solutions Tailored for

Patient-centric personalized healthcare will drive the future of wearable technology. Using AI algorithms that analyze our individual health data history over time, these wearables will be able to provide personalized recommendations based on what suits your particular state of health. This customization will increase patient engagement and help them stick to their treatment plans.

The destiny of wearable healthcare is not just founded on technology but also user behavior and clinical incorporation. A suggested Wearable Technology Adoption Model (WTAM) incorporates three crucial factors:

Behavioral Psychology—How patient participation and confidence impact adoption.

Technological Readiness—AI precision, battery capability, and data protection's role.

Clinical Effectiveness—How wearables affect patient outcomes from empirical evidence.

Using the WTAM framework, healthcare researchers can examine actual-world adoption patterns and detect impediments to more extensive use [65].

Integration with Telehealth Services

The combination of wearable technology and telehealth services will augment the broader impact of such services on health delivery. Wearables can generate real-time data during virtual consultations, later allowing healthcare providers to rely on available data about the patient's current condition when making decisions. This will form a part of the extensive management of care that wearables and telehealth promise in synergies.

Concentration on Preventive Health

Wearable devices would certainly promote preventive health effectively. Continuous monitoring of the body's vital signs coupled with alerts on health threats would lead to an individual's proactive health-maintaining behavior. Preventive health is a much larger picture of value-based care as compared to only patient outcomes in volume.

Investments in Research and Development

Investments in research and development will catalyze innovations in wearable technologies. The joint effort of an academic, industrial, and health organization will help develop more complex new devices. The enhancement in sensor technologies coupled with analytic and evaluation components, as well as a user interface in ongoing research, will make wearables more functional [66].

Global Health Initiatives

Wearable technology has a great role to play in transforming and advancing the global health initiatives directed toward effective public health outcomes. They can help provide affordable monitoring solutions to serve underprivileged segments of the population for the improvement of inequities in access to healthcare services. Wearable technologies in resource-poor settings are empowering individuals to live better lives [67, 68].

10 Emerging Trends and the Future of Wearables

Wearable technology is moving so fast that inventive applications will reshape many facets of healthcare, fitness, entertainment, and daily life.

Wearable Technology Current Landscape

The wearable technology market is booming and is anticipated to touch nearly \$186.14 billion by 2030, with a compound annual growth rate (CAGR) of 14.6% during the forecast period of 2023–2030. Development in sensor technologies, artificial intelligence (AI), and increasing demand from consumers about health monitoring devices is triggering this growth. Seamless integration of wearable technologies within everyday life makes the best user experience with added application scope [69].

Currently Developing Trends in Wearable Technology

AI Wearables

At present AI is the forerunner of wearable technology innovation. AI would assist wearables in analyzing real-time data, and would give personalized recommendations to the user regarding fitness, health management, and productivity. For example, devices such as Apple Watch Series 9 are expected to include non-invasive blood glucose monitoring, an essential factor whereby a diabetic could manage their diabetic condition without repeating blood tests. This empowers the user to take healthy steps for better health and provides the necessary data to healthcare professionals, assisting in further diagnosis and treatment.

AI driven wearables are in transition phase from passive trackers to proactive intelligent assistants predicting user preferences and behavior. This change improves user satisfaction and engagement, increasing wearables value as tools for lifestyle optimization and health management.

Augmented Reality (AR) Integration

It is among the wearables that augmented reality has been integrated, with smart glasses standing out as one of its most compelling inclusions. The next-gen AR glasses in the future offer some breathtaking experiences that blend digital information seamlessly with real-world. It goes far beyond the AR applications in wearables when it comes to entertainment enhanced navigation systems, educational tools or productivity applications for more casual enhancing tasks. As AR technology evolve, it will probably redefine environment interaction in the most fundamental way [70].

Advanced Sensor Technologies

In wearables there are some advanced sensor technologies under development. The new sensor will be able to track physiological metrics such as heart rate, blood oxygen levels and hydration, even stress. Advances and increasing exposure to sensor data mean wearables can now deliver health insights that were previously impossible.

New developments in sensor technology are going to include printed and quantum sensors which are cutting-edge instruments that use the properties of quantum mechanics to measure physical quantities with previously unheard-of accuracy and possibly with higher sensitivity than traditional sensors that will only further increase the accuracy of performance. Consider, for example, the smart clothing with sensors that track body movements and temperatures around-the-clock to furnish critical insights to fitness enthusiasts or healthcare providers [71].

IoT (Internet of Things) Connectivity

The Internet of Things (IoT) is changing how wearable devices communicate with other smart devices The wearables which are enabled by IoT can communicate with smartphones, smart home devices as well as healthcare systems in real time. Such connectivity will enable smart wearables to share data in ways that exist beyond the present confines.

For instance, user data from the fitness trackers enabled with IoT can be synchronized with health applications on smartphones which then provide a holistic perspective of one's metrics to its consumer. The resulting interconnectedness not only enhances user experience, but also enables better care coordination between Health Care Providers [72].

Focus on User Experience (UX)

Wearable technology should evolve to focus on user experience (UX) as the mainstream. The manufacturers are building and creating more user friendly and very much required overhauls of UI design. From tidying up the navigation experience on its app, to making sure wearables are comfortable for extended use.

Sustainable Wearable design

Sustainability is part of the development trend for wearable technology. With consumers becoming increasingly environmentally minded, manufacturers are looking towards the use of biodegradable materials and sustainable practices on production lines. This includes everything from making the casework of such devices out of recyclable materials to optimizing components so they consume less power.

Sustainable design not only matters to the climate conscious consumer, it also achieves at a higher level when considering wider industry objectives around cutting waste and mitigating environmental damage. In future, sustainability may offer a competitive advantage for companies [73].

Smart Clothes

Smart clothing is a cutting-edge subset of wearable technology in which textiles are fitted with sensors that can track a variety of biometric information. Real-time monitoring of body temperature, activity levels, heart rate, and other parameters is possible with these clothes.

Smart apparel provides individualized insights into health metrics and features like posture correction and haptic feedback for navigation. This invention has important ramifications for both people managing chronic illnesses and athletes trying to maximize performance.

Data Ownership

As the debate on data privacy intensifies, there is a significant rise towards shared ownership of data in terms of health wearables. They cannot possibly use any aspect of their personal health data with third parties without their approval.

Therefore, manufacturers are coming up with frameworks that allow these users to decide on the data they are sharing with either an app or a healthcare provider. This is both convenient to enhance the trust of users but also empowering users in their own context on health data [74].

Contactless Payment Facilities

The increasing introduction of contactless payment features in wearable devices has aroused a lot of interest in consumers, who find this approach to transactions very convenient for them. Wearable devices have the added functionality of Near Field Communication (NFC), which is a short-range wireless technology that makes it easy and safe for electronic devices to communicate with one another which means that payments can now be initiated and completed with the sole use of the user's smartwatch or fitness tracker.

Because it streamlines payment procedures and improves user convenience, this trend has an impact on a number of industries, including retail and transportation. It is anticipated that more customers will accept wearables as multipurpose tools as contactless payment adoption rises globally [75].

Health Status-metering

The primary concern driving innovations in wearable technology today remains health monitoring. Innovative equipment continues to develop, ranging from glucose detectors to hydration monitors, from respiratory rate monitors to sleep quality detectors. This trend generally comes with a much broader shift toward preventive health strategies that lead to early detection and remedy.

Wearable technology with sophisticated health monitoring features can give users ongoing information about their health, enabling them to make wise lifestyle decisions or seek medical help when needed [76].

11 Challenges Ahead

However, promising as wearable technology is, there are still many obstacles Since wearable have access to personal data such health information, privacy of health data is becoming primary concern. It means having strong security measures in place to retain user trust. The data collected should be accurate for the effectiveness of the wearables. Wearable technology provides numerous data streams that must be integrated and validated through repeated tests in order to be effective [77].

12 Conclusion

Wearable technology is poised to revolutionize the delivery of healthcare by providing real-time monitoring, predictive analysis, and targeted interventions. This chapter has explored the multidimensional nature of wearable devices in healthcare, from their technological underpinnings to their clinical uses and social implications. The empirical evidence presented in this chapter illustrates that wearable devices have tremendous advantages along the healthcare continuum. Clinical verification tests have established their increasing precision and reliability, with certain devices proving more than 90% accurate in tracking vital health parameters against the test criteria of medical-grade equipment. The fusion of artificial intelligence and wearable sensors is a paradigm change from reactive to preventive care. Such predictive ability, coupled with real-time monitoring, allows early interventions that can avoid complications and enhance outcomes. Nevertheless, broad-based implementation of wearable technology encounters serious problems to be overcome systemically. Issues of data privacy are top concerns, Standardization of verification of wearable health metrics is needed, because just 27% of devices available for sale are in line with acceptable standards of. Of most concern, perhaps, is unequal access to these devices, risking a further increase in present disparities in health along socioeconomic, educational, and geographical lines.

13 Key Takeaways

Wearable technology has evolved from simple tracking devices to AI-driven health management devices. Remote patient monitoring (RPM) and predictive analytics have enhanced early disease diagnosis and lowered hospital readmissions. Compliance with regulations, data protection, and standardization issues need to be overcome to achieve large-scale adoption.

14 Recommendations and Future Directions

Different recommendations include for different professionals, for healthcare professionals it includes healthcare professionals incorporate AI-enabled wearables in clinical practice for the early detection of diseases and the management of chronic diseases while engaging patients. For regulators, this includes developing uniform global policies for wearable healthcare devices, with an emphasis on data protection and clinical verification. For technology developers, this includes improving wearable device precision, battery life, and interoperability to enable seamless integration with healthcare systems.

New technologies such as sensor-integrated fabrics and biocompatible implants will provide unobtrusive, continuous health monitoring. Virtual patient models derived from real-time wearable data will advance predictive diagnostics and tailored treatment plans.5G-Enabled Wearable Networks increased connectivity will provide instant data transmission, enhancing telemedicine and remote surgery assistance. Blockchain for Wearable Data Security and decentralization of storage of health data will increase security and patient ownership of health records.

References

1. Ometov, A., Shubina, V., Klus, L., Kleineberg, J.E., Olisov, I., Lohan, E.S., et al.: A survey on wearable technology: history, state-of-the-art and current challenges. Comput. Netw. **193**, 108074 (2021). https://doi.org/10.1016/j.comnet.2021.108074
2. Liu, C., Yang, M., Di, J., Yang, L., Wu, H., Wang, X., et al.: Wearable ECG: history, key technologies and future challenges. Chin. J. Biomed. Eng. **38**(6), 641–649 (2019). https://doi.org/10.3969/j.issn.0258-8021.2019.06.001
3. Ahmed, A., Aziz, S., Alzubaidi, M., Alhujaili, A., Aslanpour, Z., Siddiqi, S., et al.: Wearable devices for anxiety and depression: a scoping review. Comput. Methods Programs Biomed. Update **3**, 100095 (2023). https://doi.org/10.1016/j.cmpbup.2023.100095
4. Iqbal, S.M.A., Mahgoub, I., Du, E., Leavitt, M.A., Asghar, W.: Advances in healthcare wearable devices. NPJ Flex. Electron. **5**(1), 9 (2021). https://doi.org/10.1038/s41528-021-00107-x
5. Nelson, E.C., Verhagen, T., Vollenbroek-Hutten, M., Noordzij, M.L.: Is wearable technology becoming part of us? Developing and validating a measurement scale for wearable technology embodiment. JMIR mHealth uHealth. **7**(8), e12771 (2019). https://doi.org/10.2196/12771
6. Lu, L., Zhang, J., Xie, Y., Gao, F., Xu, S., Wu, X., et al.: Wearable health devices in health care: narrative systematic review. JMIR mHealth uHealth **8**(11), e18907 (2020). https://doi.org/10.2196/18907
7. Escobar-Linero, E., Muñoz-Saavedra, L., Luna-Perejón, F., Domínguez-Morales, M.: Wearable health devices for diagnosis support: evolution and future tendencies. Sensors **23**(3), 1678 (2023). https://doi.org/10.3390/s23031678
8. Rahmani, A.M., Szu-Han, W., Yu-Hsuan, K., Lee, J.H., Sarrafzadeh, M., Jafari, R., et al.: The internet of things for applications in wearable technology. IEEE Access. **10**, 27568–27586 (2022). https://doi.org/10.1109/ACCESS.2022.3224487
9. Surantha, N., Atmaja, P., Wibowo, D.A.: A review of wearable internet-of-things device for healthcare. Procedia Comput. Sci. **179**, 939–943 (2021). https://doi.org/10.1016/j.procs.2021.01.080
10. Haddad, T.C., Coffey, J.D., Deng, Y., Glasgow, A.E., Christopherson, L.A., Sangaralingham, L.R., et al.: Impact of a high-risk, ambulatory COVID-19 remote patient monitoring program on utilization, cost, and mortality. MayoClinProc. **97**(8), 1460–1470 (2022). https://doi.org/10.1016/j.mayocp.2022.08.015
11. Kamei, T., Kanamori, T., Yamamoto, Y., Edirippulige, S.: The use of wearable devices in chronic disease management to enhance adherence and improve telehealth outcomes: a systematic review and meta-analysis. J. Telemed. Telecare **28**(3), 163–173 (2022). https://doi.org/10.1177/1357633X20937573
12. Tran, V.T., Riveros, C., Ravaud, P.: Patients views of wearable devices and AI in healthcare: findings from the ComPaRe e-cohort. NPJ Digit. Med. **2**, 53 (2019). https://doi.org/10.1038/s41746-019-0132-y
13. Taylor, M.L., Thomas, E.E., Snoswell, C.L., Smith, A.C., Caffery, L.J.: Does remote patient monitoring reduce acute care use? A systematic review. BMJ Open **11**(3), e040232 (2021). https://doi.org/10.1136/bmjopen-2020-040232
14. Lee, J., Kim, D., Ryoo, H.Y., Shin, B.S.: Sustainable wearables: wearable technology for enhancing the quality of human life. Sustainability **8**(5), 466 (2016). https://doi.org/10.3390/su8050466
15. Patel, V., Chesmore, A., Legner, C.M., Pandey, G.: Trends in workplace wearable technologies and connected-worker solutions for next-generation occupational safety, health, and productivity. Adv. Intell. Syst. **4**(5), 2100099 (2022). https://doi.org/10.1002/aisy.202100099
16. Seshadri, D.R., Li, R.T., Voos, J.E., Rowbottom, J.R., Alfes, C.M., Zorman, C.A., et al.: Wearable sensors for monitoring the internal and external workload of the athlete. NPJ Digit. Med. **2**, 71 (2019). https://doi.org/10.1038/s41746-019-0149-2
17. Shei, R.J., Holder, I.G., Oumsang, A.S., Paris, H.L., Stickford, J.L.: Wearable activity trackers—advanced technology or advanced marketing? Eur. J. Appl. Physiol. **122**, 2505–2516 (2022). https://doi.org/10.1007/s00421-022-04951-1

18. Mardonova, M., Choi, Y.: Review of wearable device technology and its applications to the mining industry. Energies **11**(3), 547 (2018). https://doi.org/10.3390/en11030547
19. Min, S., Kim, D.H., Joe, D.J., Bae, S., Lee, M., Yoo, S., et al.: Clinical validation of a wearable piezoelectric blood-pressure sensor for continuous health monitoring. Adv. Mater. **35**(13), 2301627 (2023). https://doi.org/10.1002/adma.202301627
20. Ates, H.C., Nguyen, P.Q., Gonzalez-Macia, L., Morales-Narváez, E., Güder, F., Dincer, C., et al.: End-to-end design of wearable sensors. Nat. Rev. Mater. **7**, 887–907 (2022). https://doi.org/10.1038/s41578-022-00460-x
21. Shumba, A.T., Montanaro, T., Sergi, I., Bramanti, A., Ciccarelli, M., Rispoli, A., et al.: Wearable technologies and AI at the far edge for chronic heart failure prevention and management: a systematic review and prospects. Sensors **23**(15), 6896 (2023). https://doi.org/10.3390/s23156896
22. Jin, C.Y.: A review of AI technologies for wearable devices. In: IOP Conference Series: Materials Science and Engineering, vol. 688, p. 044072. IOP Publishing (2019). https://doi.org/10.1088/1757-899X/688/4/044072.
23. Celic, L., Magjarevic, R.: Seamless connectivity architecture and methods for IoT and wearable devices. Automatika. **61**(3), 333–340 (2020). https://doi.org/10.1080/00051144.2019.1660036
24. Guo, Y., Liu, X., Peng, S., Jiang, D., Xu, K., Xiao, M., et al.: A review of wearable and unobtrusive sensing technologies for chronic disease management. Comput. Biol. Med. **129**, 104163 (2021). https://doi.org/10.1016/j.compbiomed.2020.104163
25. Yu, S., Chen, Z., Wu, X.: The impact of wearable devices on physical activity for chronic disease patients: findings from the 2019 health information national trends survey. Int. J. Environ. Res. Public Health **20**(1), 887 (2023). https://doi.org/10.3390/ijerph20010887
26. Hrabovska, N., Kajati, E., Zolotova, I.: A validation study to confirm the accuracy of wearable devices based on health data analysis. Electronics **12**(11), 2536 (2023). https://doi.org/10.3390/electronics12112536
27. Miller, D.J., Sargent, C., Roach, G.D.: A validation of six wearable devices for estimating sleep, heart rate and heart rate variability in healthy adults. Sensors **22**(16), 6317 (2022). https://doi.org/10.3390/s22166317
28. Guk, K., Han, G., Lim, J., Jeong, K., Kang, T., Lim, E.K., et al.: Evolution of wearable devices with real-time disease monitoring for personalized healthcare. Nanomaterials **9**(6), 813 (2019). https://doi.org/10.3390/nano9060813
29. Dias, D., Cunha, J.P.S.: Wearable health devices—vital sign monitoring, systems and technologies. Sensors **18**(8), 2414 (2018). https://doi.org/10.3390/s18082414
30. Versteeg, H., Pedersen, S.S., Mastenbroek, M.H., Redekop, W.K., Schwartz, P.J., Mabo, P., et al.: Patient perspective on remote monitoring of cardiovascular implantable electronic devices: rationale and design of the REMOTE-CIED study. Neth Heart J. **22**(10), 423–428 (2014). https://doi.org/10.1007/s12471-014-0587-z
31. Xie, Y., Lu, L., Gao, F., Yang, H., Liu, Y., Li, Y., et al.: Integration of artificial intelligence, blockchain, and wearable technology for chronic disease management: a new paradigm in smart healthcare. Curr. Med. Sci. **41**(6), 1108–1115 (2021). https://doi.org/10.1007/s11596-021-2485-0
32. Giurgiu, M., Ketelhut, S., Kubica, C., Nissen, R., Doster, A.K., Thron, M., et al.: Assessment of 24-hour physical behaviour in adults via wearables: a systematic review of validation studies under laboratory conditions. Int. J. Behav. Nutr. Phys. Act. **20**(1), 73 (2023). https://doi.org/10.1186/s12966-023-01473-7
33. Yeh, C.Y., Chung, Y.T., Chuang, K.T., Chen, L.C., Lee, M.H., Huang, C.C., et al.: An innovative wearable device for monitoring continuous body surface temperature (HEARThermo): instrument validation study. JMIR mHealth uHealth. **9**(7), e19210 (2021). https://doi.org/10.2196/19210
34. Spagnolli, A., Guardigli, E., Orso, V., Varotto, A., Gamberini, L.: Measuring user acceptance of wearable symbiotic devices: validation study across application scenarios. In: Jacucci, G., Gamberini, L., Freeman, J., Spagnolli, A., (eds). Symbiotic Interaction. Lecture Notes in Computer Science, vol. 8820, pp. 87–98. Springer, Cham (2014). https://doi.org/10.1007/978-3-319-13500-7_7

35. Lewy, H.: Wearable technologies-future challenges for implementation in healthcare services. Healthc Technol Lett. **2**(1), 2–5 (2015). https://doi.org/10.1049/htl.2014.0104
36. Cardinale, M., Varley, M.C.: Wearable training-monitoring technology: applications, challenges, and opportunities. Int. J. Sports Physiol. Perform. **12**(S2), S255–S262 (2017). https://doi.org/10.1123/ijspp.2016-0423
37. Hernandez, N., Lundström, J., Favela, J., McChesney, I., Arnrich, B.: Literature review on transfer learning for human activity recognition using mobile and wearable devices with environmental technology. SN Comput. Sci. **1**, 66 (2020). https://doi.org/10.1007/s42979-020-0070-4
38. Cahyanto, I.: Privacy challenges in using wearable technology in education literature review. Formosa. J. Appl. Sci. **2**(6), 156–164 (2023). https://doi.org/10.55927/fjas.v2i6.4272
39. Liao, Y., Thompson, C., Peterson, S., Mandrola, J., Beg, M.S.: The future of wearable technologies and remote monitoring in health care. Am. Soc. Clin. Oncol. Educ. Book **39**, 115–121 (2019). https://doi.org/10.1200/edbk_238919
40. Krummel, T.M.: The rise of wearable technology in health care. JAMA Netw. Open **2**(2), e187672 (2019). https://doi.org/10.1001/jamanetworkopen.2018.7672
41. Lymberis, A.: Progress in R&D on wearable and implantable biomedical sensors for better healthcare and medicine. In: Proceedings of the 3rd IEEE/EMBS Special Topic Conference on Microtechnology in Medicine and Biology; 2005 May 12–15; Oahu, Hawaii, pp. 294–296. IEEE (2005). https://doi.org/10.1109/MMB.2005.1548453
42. Chen, Y.H., Sawan, M.: Trends and challenges of wearable multimodal technologies for stroke risk prediction. Sensors **21**(2), 460 (2021). https://doi.org/10.3390/s21020460
43. Vijayan, V., Connolly, J., Condell, J., McKelvey, N., Gardiner, P.: Review of wearable devices and data collection considerations for connected health. Sensors **21**(16), 5589 (2021). https://doi.org/10.3390/s21165589
44. An, J.: Wearable device-based data collection and feature analysis method for outdoor sports. Int. J. Distrib. Syst. Technol. **13**(3), 1–12 (2022). https://doi.org/10.4018/IJDST.307992
45. Nayak, B., Bhattacharyya, S.S., Krishnamoorthy, B.: Integrating wearable technology products and big data analytics in business strategy: a study of health insurance firms. J. Syst. Inf. Technol. **21**(5), 535–550 (2019). https://doi.org/10.1108/JSIT-08-2018-0109
46. Bouderhem, R.: Privacy and regulatory issues in wearable health technology. Eng. Proc. **28**(1), 46 (2023). https://doi.org/10.3390/ecsa-10-16206
47. Jiang, N., Mück, J.E., Yetisen, A.K.: The regulation of wearable medical devices. Trends Biotechnol. **38**(2), 129–133 (2020). https://doi.org/10.1016/j.tibtech.2019.06.004
48. Wakabayashi, Y., Matsui, H., Ikai, K., Hasegawa, N., Miyano, M., Watanabe, T., et al.: Developing a practical method for validation of computerized systems integrated with smart and/or wearable devices for regulatory compliance of clinical trials. Ther. Innov. Regul. Sci. **51**(6), 780–786 (2017). https://doi.org/10.1177/2168479016666585
49. Akshay, A.M., Venkatesh, M.P., Kumar, P.T.M.: Wearable healthcare technology—the regulatory perspective. Int. J. Drug. Regul. Aff. **4**(1), 29–36 (2016). https://doi.org/10.22270/ijdra.v4i1.13
50. Jung, C.K., Kim, J., Rhim, H.: Validation of an ear-worn wearable gait analysis device. Sensors **23**(3), 1244 (2023). https://doi.org/10.3390/s23031244
51. McNaboe, R., Beardslee, L., Kong, Y., Daley, R., Rosenfeld, D., Gu, Y., et al.: Design and validation of a multimodal wearable device for simultaneous collection of electrocardiogram, electromyogram, and electrodermal activity. Sensors **22**(22), 8851 (2022). https://doi.org/10.3390/s22228851
52. Birkeland, K., Khandwalla, R.M., Kedan, I., Shufelt, C.L., Mehta, P.K., Minissian, M.B., et al.: Daily activity measured with wearable technology as a novel measurement of treatment effect in patients with coronary microvascular dysfunction: substudy of a randomized controlled crossover trial. JMIR Res. Protoc. **6**(12), e255 (2017). https://doi.org/10.2196/resprot.8057
53. Izmailova, E.S., McLean, I.L., Bhatia, G., Hather, G., Cantor, M., Merberg, D., et al.: Evaluation of wearable digital devices in a phase I clinical trial. Clin. Transl. Sci. **12**(3), 247–256 (2019). https://doi.org/10.1111/cts.12602

54. Colvonen, P.J., De Young, P.N., Bosompra, N.O.A., Owens, R.L.: Limiting racial disparities and bias for wearable devices in health science research. Sleep. **43**(10), zsaa159 (2020). https://doi.org/10.1093/sleep/zsaa159
55. Dhingra, L.S., Aminorroaya, A., Oikonomou, E.K., Mahajan, S., Nasir, K., Khera, R.: Use of wearable devices in individuals with or at risk for cardiovascular disease in the US, 2019 to 2020. JAMA Netw. Open **6**(5), e2316634 (2023). https://doi.org/10.1001/jamanetworkopen.2023.16634
56. Keshet, A., Reicher, L., Bar, N., Thaiss, C.A.: Wearable and digital devices to monitor and treat metabolic diseases. Nat. Metab. **5**(6), 759–774 (2023). https://doi.org/10.1038/s42255-023-00778-y
57. Schuster, A.M., Kadylak, T., Cotten, S.R.: Correlation between socio-demographic factors and adoption and use of wearable activity trackers in online American older adults. Educ. Gerontol. **49**(3), 195–206 (2023). https://doi.org/10.1080/03601277.2022.2064688
58. Leroux, A., Rzasa-Lynn, R., Crainiceanu, C., Sharma, C., Walia, A., Knoerl, R., et al.: Wearable devices: current status and opportunities in pain assessment and management. Digit. Biomark. **5**(1), 89–102 (2021). https://doi.org/10.1159/000515576
59. Beg, S., Handa, M., Shukla, R., Kaur, R., Jain, S., Imam, S.S., et al.: Wearable smart devices in cancer diagnosis and remote clinical trial monitoring: transforming the healthcare applications. Drug Discov. Today **27**(8), 2307–2315 (2022). https://doi.org/10.1016/j.drudis.2022.06.014
60. Sharma, A., Gupta, R., Singh, M., Kumar, V., Patel, R., Desai, N., et al.: Transition to value-based care: IoT wearable devices in maternity healthcare for low-resource countries. Int. J. Sci. Res. Comput. Sci. Eng. **11**(4), 8–14 (2023). https://doi.org/10.26438/ijsrcse/v11i4.814
61. Nedunchezhian, S., Dagher, L., Russell, A., Aung, K., Drennan, K., Yelamanchili, V.S.: B-PO02-182 disparities in wearable health device and telemedicine understanding and utilization amongst cardiology patients in New Orleans. Heart Rhythm **18**(8), S233 (2021). https://doi.org/10.1016/j.hrthm.2021.06.435
62. Akam, E.Y., Nuako, A.A., Daniel, A.K., Adejumo, A.C., Feyissa, T.M.: Racial disparities and cardiometabolic risk: new horizons of intervention and prevention. Curr. Diab. Rep. **22**(10), 433–444 (2022). https://doi.org/10.1007/s11892-022-01451-6
63. Cox, S.M., Lane, A., Volchenboum, S.L.: Use of wearable, mobile, and sensor technology in cancer clinical trials. JCO Clin. Cancer Inform. **2**, 1–11 (2018). https://doi.org/10.1200/cci.17.00147
64. Onyeaka, H., Muoghalu, C., Malekani, M., Fogleman, C., Sterns, A., Fogleman, M.: Use of wearable devices among individuals with depression and anxiety: a population level study. Psychiatry Res. Commun. **2**(1), 100081 (2022). https://doi.org/10.1016/j.psycom.2022.100081
65. Alduaij, M.Y.: Towards a wearable technology model: implications for wearable technology adoption. Int. J. Inf. Syst. Serv. Sect. **14**(1), 1–17 (2022). https://doi.org/10.4018/IJISSS.295869
66. Ahn, C.R., Lee, S., Sun, C., Jebelli, H., Yang, K., Choi, B.: Wearable sensing technology applications in construction safety and health. J. Constr. Eng. Manag. **145**(11), 03119007 (2019). https://doi.org/10.1061/(asce)co.1943-7862.0001708
67. The future toward: interactivity in wearable devices. Front. Soc. Sci. Technol. **4**(5), 1–9 (2022). https://doi.org/10.25236/fsst.2022.040508
68. Mäkynen, M., Ng, G.A., Li, X., Schlindwein, F.S., Tuan, J.H., Eftekhari, A., et al.: Wearable devices combined with artificial intelligence—a future technology for atrial fibrillation detection? Sensors **22**(22), 8588 (2022). https://doi.org/10.3390/s22228588
69. Çinier, G.: Wearable technologies in cardiology: current evidence and future perspective. Anatol. J. Cardiol. **22**(Suppl 2), 21–25 (2019). https://doi.org/10.14744/AnatolJCardiol.2019.55550
70. Lee, S.M., Lee, D.H.: Healthcare wearable devices: an analysis of key factors for continuous use intention. Serv. Bus. **14**, 503–531 (2020). https://doi.org/10.1007/s11628-020-00428-3
71. Kodithuwakku Arachchige, S.N.K., Burch, V.R.F., Chander, H., Knight, A., Cox, C.A., Luczak, T., et al.: The use of wearable devices in cognitive fatigue: current trends and future intentions. Theor. Issues Ergon. Sci. **23**(2), 244–267 (2022). https://doi.org/10.1080/1463922X.2021.1965670

72. Byrne, J.R., O'Sullivan, K., Sullivan, K.: An IoT and wearable technology hackathon for promoting careers in computer science. IEEE Trans. Educ. **60**(1), 50–58 (2017). https://doi.org/10.1109/TE.2016.2626252
73. Loncar-Turukalo, T., Zdravevski, E., da Silva, J.M., Chouvarda, I., Trajkovik, V.: Literature on wearable technology for connected health: scoping review of research trends, advances, and barriers. J. Med. Internet Res. **21**(9), e14017 (2019). https://doi.org/10.2196/14017
74. Miao, F., Wu, D., Liu, Z., Jiang, Z., Yan, Y., Cao, Y., et al.: Wearable sensing, big data technology for cardiovascular healthcare: current status and future prospective. Chin. Med. J. **136**(7), 772–780 (2023). https://doi.org/10.1097/CM9.0000000000002117
75. Beniczky, S., Karoly, P., Nurse, E., Ryvlin, P., Cook, M.: Machine learning and wearable devices of the future. Epilepsia **62**(S2), S116–S124 (2021). https://doi.org/10.1111/epi.16555
76. Kim, H., Pyun, K.R., Lee, M.T., Yang, H., Kang, J.M., Kwon, O.S., et al.: Recent advances in sustainable wearable energy devices with nanoscale materials and macroscale structures. Adv. Funct. Mater. **32**(12), 2110535 (2022). https://doi.org/10.1002/adfm.202110535
77. Sharma, V., Vineeta, Som, S., Jain, S., Sharma, A.: Future of wearable devices using IoT synergy in AI. In: Proceedings of the 3rd International Conference on Electronics and Communication and Aerospace Technology (ICECA 2019); 2019 Jun 12–14; Coimbatore, India, pp. 1204–1208. IEEE (2019). https://doi.org/10.1109/ICECA.2019.8821915

Vikrant Verma is a Professor at Kharvel Subharti College of Pharmacy, Swami Vivekanand Subharti University, Meerut, India. With over 14 years of experience in teaching and research, Dr. Verma's expertise lies in pharmaceutical chemistry, specifically in drug synthesis and characterization. He holds a Ph.D. in Pharmaceutical Chemistry and has authored over 60 peer-reviewed publications indexed in Scopus and Web of Science. He is a recipient of multiple research grants and holds numerous patents, both Indian and international. Dr. Verma has also written over 15 scholarly books and serves on the editorial and advisory boards of reputed journals.

Charu Saxena is an Associate Professor at Kharvel Subharti College of Pharmacy with more than a decade of academic experience. She earned her M.Pharm in Pharmaceutics and her Ph.D. in Pharmaceutical Sciences. Her research interests include novel drug delivery systems, dosage form development, and solubility enhancement techniques. Dr. Saxena has published over 30 articles in Scopus and Web of Science indexed journals and authored over six books. She holds several national and international patents and has successfully mentored more than six postgraduate students. She also contributes as a research guide and serves on various academic panels.

The Role of Internet of Things (IoT) in Wearable Devices for Disabled Individuals

Md. Anisul Islam, Radhika Ramachandran Nair, Gadha Jyothi Prasad, and Rafiqul Chowdhury

Abstract The wearable Internet of Things (IoT) holds great promise for individuals with disabilities, offering the potential to enhance their participation in society and the workforce. To maximize these benefits, it is essential to thoughtfully address the challenges and limitations associated with integrating IoT into wearable healthcare devices. In this chapter, we present a comprehensive and effective architecture for wearable IoT that organizes and manages patient data efficiently. This approach ensures well-coordinated medical services and timely care tailored specifically for individuals with disabilities. We also explore a range of innovative wearable IoT solutions designed to address various challenges faced by these individuals, such as accessibility barriers, financial insecurity, social exclusion, educational disparities, healthcare inequities, and legal and transportation obstacles. Additionally, we discuss the technical, ethical, and implementation challenges that must be addressed to create effective wearable IoT solutions. By identifying these opportunities and challenges, we can work towards developing more inclusive solutions that incorporate machine learning techniques, advanced data analytics, input from practitioners, and blockchain technology within wearable IoT structures. These cross-sector efforts have the potential to significantly empower individuals with disabilities, fostering a more inclusive and supportive environment for everyone involved in healthcare systems.

Keywords Internet of Things (IoT) · Wearable devices · Disabled individuals · Internet of Medical Things (IoMT)

Md. A. Islam (✉) · R. R. Nair · G. J. Prasad · R. Chowdhury
Department of Management Science, Shannon School of Business, Cape Breton University, Sydney, NS, Canada
e-mail: Mdanisul_Islam@cbu.ca

P. Eappen et al. (eds.), *Remote Monitoring and Wearable Devices in Healthcare*, Information Systems Engineering and Management 63,
https://doi.org/10.1007/978-3-031-98897-4_12

1 Introduction

1.1 Internet of Things (IoT) in Healthcare System

In the era of rapid advancement in digital technology, notably through the ongoing evolution of the internet and mobile devices, the healthcare system is experiencing significant and transformative improvement. These innovations pave the way for a smart healthcare system, and disabled individuals, where integration of real-time data analysis and processing is crucial for enhancing efficient medical services for disabled and general patients and all stakeholders involved.

The Internet of Things (IoT) involves technologies that connect both wire and wireless physical devices equipped with sensors and actuators through internet protocols. This allows devices to exchange and store data efficiently across the network without or with minimal human intervention. In the medical field, the concept of IoT is referred to as the Internet of Medical Things (IoMT). The IoMT combines a range of smart devices such as wearables (like smartwatches, smart glasses, smart clothing, ECG sensors, and smart contact lenses) with hospital resources and medical examination tools to foster communication and empower informed decision-making, primarily on a real-time basis.

1.2 Wearable Devices for Disabled Individuals

Wearable devices are electronics smart devices designed to be attached or worn on the human body, utilize their in-built electronic and biosensors to monitor physical, biological and physiological signals. Since its landmark redevelopment in the year of early 2000s, the devices become a game-changer in the healthcare domain [1].

Modern wearable devices sense all basic human senses including touch (skin, fin-gernails, and hair), vision (eyes), hearing, taste, taste, and proprioception (kinesthesia- human body's ability to sense muscle movement, action, and location). The devices can efficiently measure physical metrics including heart rate, water levels, sleep pat-tern, blood pressure, body temperature, physical activity (steps taken), calories burned, nutrition intake, physical strains, etc.; and biological metrics including emo-tions, sweat level, blood sugar, blood oxygen levels, stress level, sleep quality, sleep patterns, etc. Wearable devices are becoming famous because they are economical, smart, easily wearable, convenient, and interconnectable with other smart devices.

Wearable devices provide a promising solution for enhancing remote health monitoring, benefiting both patients and healthcare providers. These integrated remote monitoring technologies in the devices are particularly beneficial for individuals with chronic disorders, those needing long-term care, the elderly, and even newborn babies, as they provide real-time surveillance of health records.

The importance of remote monitoring through these devices is particularly evident for individuals with disabilities. Since they are designed to function autonomously with minimal space requirements and at a low cost. By facilitating continuous and real-time data sharing, these devices significantly improve the healthcare and daily life experience for all parties involved—patients, doctors, nurses, and hospitals—by generating substantial health data, ensuring constant supervision, and minimizing the need for in-person visits for patients, particularly with disabilities. Moreover, these devices are especially crucial for patients in rural areas or underprivileged communities where access to healthcare support is limited or risky, particularly during times of crisis like the COVID-19 pandemic [2].

Wearable devices are particularly impactful in this evolving healthcare landscape and vibrant work-life environments due to their ability to continuously monitor signals and identify various human activities automatically or with a minimum human effort. This capability facilitates a smooth exchange of data and transparent communication among various parties in the system, regardless of whether the users of devices are in a static position or motion.

According to the World Health Organization (WHO), there are around 1400 million individuals globally living with disabilities, each facing unique challenges and requir-ing different forms of support [3]. The economic and societal impact of excluding individuals with disabilities from mainstream society and the workforce are signifi-cant. Thus, addressing their needs can substantially reduce a country's societal and economic costs.

The wearable devices without internet connections (ultimately without the Internet of Things) work as individual assistive technology for users. For example, a user uses wearable devices- smart sticks, fitness trackers, hearing aids, smartwatch, Optacons, and many more to measure or monitor their physical activities or signals. Since de-vices are not connected to the internet, they are not capable of processing and exchanging data with other devices through the network. When a wearable device connects with the Internet of Things (internet of devices), it can automatically measure data, exchange data, and make intellectual decisions through the network. For example, the heart rate of a patient can automatically be measured by the Holter moni-tor in a hospital.

2 Internet of Things (IoT) in Wearable Devices for Disabled Individuals

2.1 Wearable Devices with IoT

Wearable devices provide stakeholders—such as patients, doctors, nurses, hospitals, working colleagues, friends, and service providers—with the ability to generate real-time data about individuals and patients in various situations. This timely diagnostic information allows for informed, data-driven decisions, ultimately ensuring that individuals or patients receive the right medical care and service care when needed. By facilitating precise and effective communication among all devices in the system, the real-time monitoring, collaboration, data storage and usage are enhanced.

The Internet of Things (IoT) plays a central role in data transmission process by connecting all parties and devices, enabling proactive oversight and integration of devices and technologies, and sharing the data, all while minimizing routing tasks, costs, and time in the healthcare or service providing system. When Internet of Things (IoT) technology is integrated into wearable devices, the concept is called "wearable IoT." This interconnected approach not only improves patient or individual' outcomes but also streamlines healthcare processes and service facilities for everyone involved.

Thus, the IoMT (i.e., Internet of Things for medical devices) presents a remarkable opportunity to improve patient care and outcomes through continuous, real-time data exchange, ultimately leading to improved response times across the healthcare system and caregiving systems. This continuous monitoring not only reduces medical costs and minimizes the risk of errors, but also optimizes the use of medical resources, resulting in lower overall healthcare expenditures. Additionally, it promotes greater patient engagement by granting access to health data, ensuring effective communication with service providers, and facilitating smoother, well-coordinated, and more affordable medical services. The features of IoMT discussed greatly enhance the mobility and accessibility of individuals, particularly those with disabilities leading to a better quality of life and increased participation in the labor market. Figure 1 illustrates a wearable Internet of Things (wearable IoT) for disabled individuals.

2.2 A Wearable Internet of Things (IoT) for Disabled Individuals

The design of wearable IoT for disabled individuals consists of four layers: sensor or data collection layer, network layer, data processing layer, and application layer as shown in Fig. 1. All layers play an important role in enhancing accessibility, mobility, and support for individuals with disabilities by leveraging smart sensors, networking technologies, data exchange systems, and intelligent smart applications.

Sensor or data collection layer: In this layer, sensors integrated into various smart devices—whether worn on the body or used in medical settings—actively gather, exchange, and process valuable data from patients or disabled persons. The collected data is then transmitted to the network level and ultimately stored centrally within the system for applications. These sensing devices function as the "eyes and ears" playing an essential role in data collection that supports effective analysis and informed decision-making. Typically, a variety of heterogeneous devices—such as different smart technologies, RFID, QR codes, GPS, and Bluetooth—work collaboratively

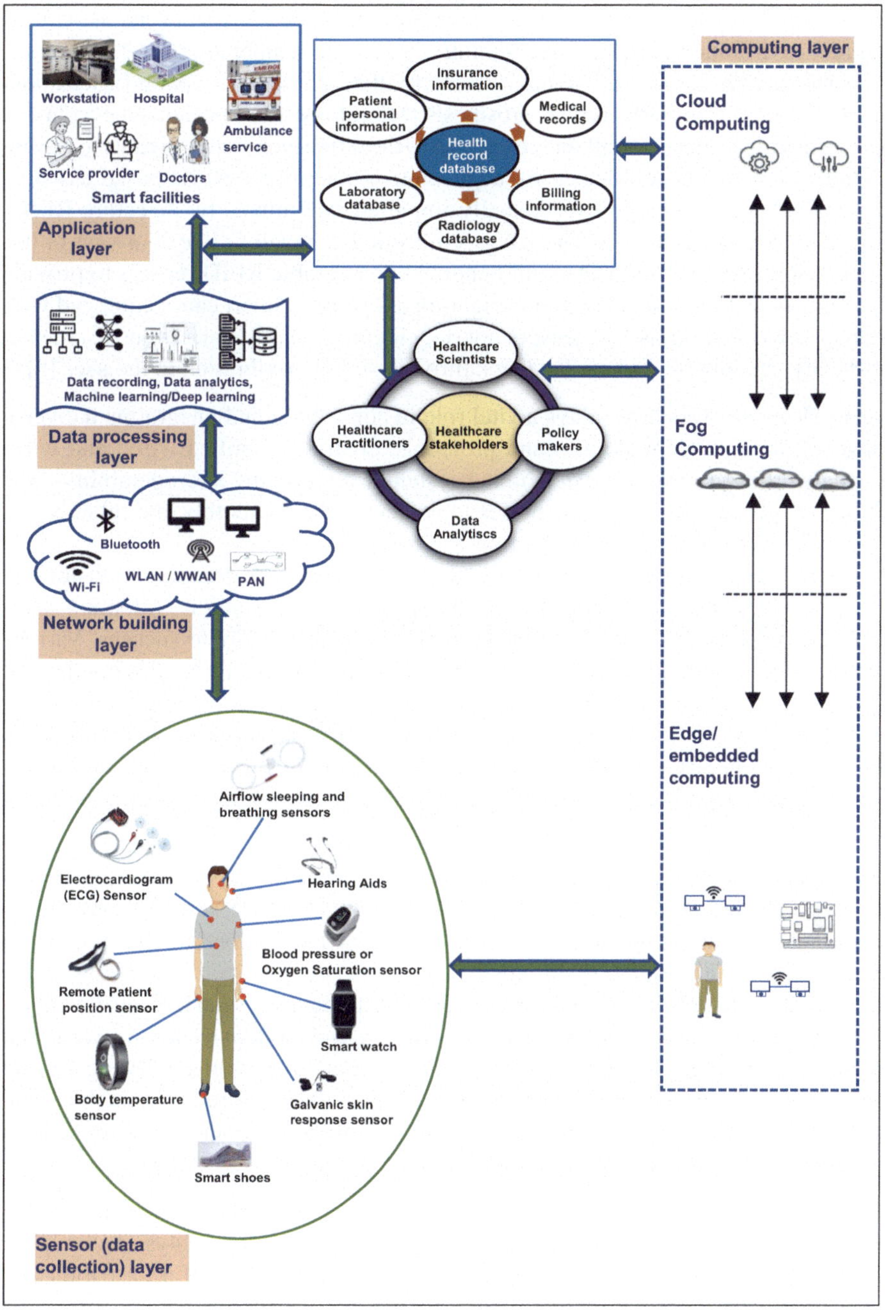

Fig. 1 Illustration a wearable internet of things (wearable IoT) for disabled individuals, inspired by study [2, 4]

within this layer to directly measure heartbeat and blood sugar levels, as well as indirect environmental factors like humidity and room temperature.

For example, a smart glove can vibrate to alert a visually impaired user about nearby hazards. Similarly, retinal prosthesis technology uses specialized eyeglasses to transmit image data to implants connected to the retina, converting signals to restore partial vision for those affected by degenerative conditions. RFID-based assistive devices enhance navigation for visually impaired individuals by detecting RFID-tagged objects or landmarks. These tags, embedded in locations like doorways or bus stops, wirelessly communicate with handheld or wearable RFID readers to provide real-time object identification and navigation assistance. Smart canes equipped with RFID readers and ultrasonic sensors not only identify nearby objects but also detect obstacles, sending immediate feedback through sound or vibration to the user [5].

Network layer: This layer plays a vital role in connecting and integrating a diverse range of devices, technologies, and protocols, creating a unified wired and wireless network for transmitting data in the system. Emphasizing interoperability—the ability of different devices and entities to exchange data smoothly and reliably—is essential for optimizing data transmission.

By ensuring seamless communication, an efficient, error-free data transfer can be achieved while minimizing the need for human intervention. Adopting standard and responsive devices and technologies is therefore key to success. Various networking technologies, such as Wi-Fi, WLAN/WWAN, PAN, Bluetooth, and Zigbee, are employed in this layer to facilitate secure and efficient data transmission. For example, RFID-enabled smart canes transmit real-time navigation data for visually impaired travelers to monitoring stations using Bluetooth or ZigBee communication protocols, enhancing user safety and confidence during mobility [5].

Data processing layer: This layer focuses on effectively processing and exchanging the data collected from the sensor layers. This involves filtering, sorting, aggregating, preparing, and validating the data to ensure seamless integration into the application layer, where all stakeholders can utilize it.

Since the sensor layers consist of a variety of heterogeneous devices and technologies, a range of data analytics methods, algorithms, and modern techniques- artificial intelligence, machine learning, and deep learning tools are employed. By identifying and analyzing key data characteristics—such as patterns, trends, anomalies, and outliers—only the most relevant and actionable information is sent to the application layers. This enhances decision-making and overall outcomes in the system.

Application layer: The top layer of the IoT architecture plays a crucial role by integrating all components, including devices, technologies, data analysis techniques, and tools. This layer serves as a central hub, delivering essential information and services to stakeholders through diverse applications within the system. This layer is where IoT systems provide meaningful services, such as personalized navigation, smart home automation, or real-time alerts tailored to user needs.

By integrating machine learning (ML), deep learning, and artificial intelligence (AI) in the data processing layer, the application layer enables advanced functionalities such as speech-to-text conversions for hearing-impaired users or predictive learning modules for adaptive environments. Navigation applications for visually impaired individuals use voice-guided directions based on sensory input to assist with safe travel. Additionally, IoT-enabled smart home solutions allow physically impaired users to control lights, appliances, and security systems using intuitive interfaces [5].

The layer enhances the overall effectiveness of the IoT architecture by ensuring that all components collaborate efficiently and safely. Notable applications in this layer include innovative facilities such as smart hospitals, ambulance services, and the vital functions performed by stakeholders—employers, service providers, doctors, nurses, and caregivers. Overall, this multi-layered IoT architecture empowers individuals with sensory, physical, or cognitive impairments, promoting greater independence, safety, and participation in everyday life through adaptive and intelligent technological solutions.

Computing layer: This layer offers the computational capabilities necessary for data processing across all devices and technologies used in different layers of the system. It encompasses all the required software and hardware support.

This layer facilitates various functions, including distributed computing, where multiple users or devices from different layers collaborate on a single decision-making process. It also supports cloud computing, fog computing, edge computing, and embedded computing. Depending on the complexity of the data processes involved, this layer can be strategically deployed either in a central cloud environment or on dedicated edge devices, optimizing performance and efficiency throughout the system.

A great example of wearable IoT technology is a chest strap that monitors a patient's heart rate (shown in Fig. 2). This device allows for real-time health tracking, which can lead to improved management of the patient's well-being.

Chest straps equipped with data monitoring devices and applications play a crucial role in accurately measuring patients' heart rates by capturing the intricate electrical activity of the heart. This innovative technology serves a wide range of

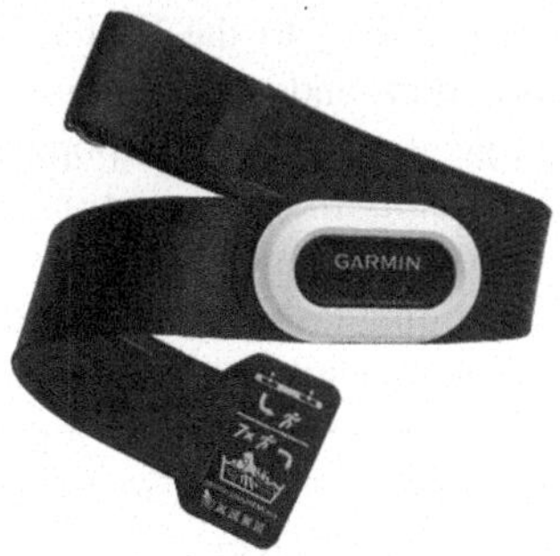

Fig. 2 A chest strap @Garmin store

patients, whether they are physically disabled, at rest, engaged in physical activity, or transitioning between different levels of exertion.

Chest straps are designed to efficiently sense and transmit heart rate data to monitoring and storage devices, such as smartwatches or fitness trackers, in the data collection layer of IoT architecture. They use Bluetooth or wired connections to ensure smooth data transfer through Wi-Fi or other reliable network protocols in the networking layer.

In the data processing layer, healthcare professionals can analyze heart rate patterns to identify irregularities that may indicate potential health concerns, such as heart attacks, strokes, or blockages. The incorporation of advanced data analytics software, algorithms, and machine learning techniques significantly enhances the accuracy of these assessments, providing valuable insights for future care.

At the application layer, heart rate data can be easily shared with healthcare stakeholders through various channels, including cloud storage, file sharing, and direct integration with healthcare software. This integration of technology and healthcare empowers doctors, nurses, hospitals, and healthcare providers to engage meaningfully with patients about their heart health. Equipped with valuable insights, they can offer personalized guidance and formulate informed recommendations for treatment or lifestyle modifications. This collaborative approach not only promotes better health outcomes but also encourages patients to take an active role in their health journey, leading to a more supportive and engaged healthcare experience.

3 Disabilities and Their Types

Disability encompasses a range of physical and mental impairments that can affect a person's ability to engage in essential daily activities and fully participate in society. Disabilities can be inherited or developed due to trauma, illness, or aging, and some are visible while others are not. Various factors influence disability, including age, gender, life stage, exposure to environmental risks, financial status, culture, and available resources. In many regions, there has been an increase in the prevalence of chronic illnesses and injuries, such as diabetes, heart disease, mental illness, cancer, and respiratory conditions, which are associated with higher rates of disability.

According to the WHO, disability has three dimensions: impairment, activity limitation, and participation restriction. Impairment is defined as deviation from the generally accepted population standard of physical form or function of body. Conditions like amputation, vision impairment, and cognitive decline are examples of impairment. Activity limitation refers to the challenges people face when performing tasks that are thought to be typical for humans, such as seeing, hearing, walking, or solving problems. The third one, participation restriction refers to limitations or restrictions on a person's involvement in life circumstances in all areas, such as employment, social contacts, leisure activities, and access to healthcare and preventive services.

Individuals with disabilities often experience significant disparities in social and economic outcomes compared to their non-disabled counterparts. These disparities primarily arise from systemic barriers and societal inequities. Disabled individuals and their families frequently incur higher expenses in various aspects of life to achieve a standard of living similar to that of those without disabilities. Additionally, having a disability nearly doubles the risk of developing conditions such as depression, asthma, diabetes, stroke, obesity, and poor dental health. These heightened risks largely result from unjust circumstances, including poverty, exclusion from education and employment, stigma, discrimination, and obstacles within the healthcare system. Table 1 provides an informative overview of various types of disabilities, along with examples that illustrate each category effectively.

Many disabilities, like those involving mobility limits, are easily observable, but many others are invisible and have a major influence on day-to-day functioning without being immediately apparent. Due to their mild nature, these conditions—which can impact mental, neurological, or bodily functions—frequently go undetected. In actuality, invisible illnesses affect a sizable portion of people with long-term medical disorders. These can include ailments that substantially interfere with day-to-day functioning, such as diabetes, renal failure, and sleep difficulties. The idea of "invisible disability" challenges our preconceptions in a world where disability is frequently connected to obvious indicators like wheelchairs, ramps, and assistive technology. People with vision or auditory problems, for instance, might not always wear glasses or hearing aids, which can make their conditions less noticeable [12].

A person's life can be greatly impacted by a variety of conditions that fall under the broad category of disabilities. These disorders can impact cognitive, sensory, emotional, and physical functioning, creating particular difficulties in day-to-day activities, social relationships, and general health. Establishing an inclusive and equitable society requires acknowledging the variety of impairments and attending to the special needs of those who have them.

4 Challenges Faced by Disabled Individuals

In their daily lives, people with disabilities face many challenges, many of which are caused by social norms and structural impediments. These difficulties may have a major effect on their independence, well-being, and standard of living. Limited autonomy, heightened susceptibility to abuse and human rights breaches, and the potential of reliance on caregivers—who may also suffer from exhaustion or inadequate support—are among the main issues. The difficulties that persons with disabilities encounter will be covered in more detail in the sections that follow.

Table 1 A brief overview of various types of disabilities along with examples

Disability types	Description with examples
Physical disabilities	These conditions can impact physical movement, endurance, and coordination, presenting unique challenges that individuals navigate daily Examples: • Impaired mobility: This includes various conditions such as paralysis from spinal cord injuries or strokes, amputations, and muscular dystrophy, which may affect mobility and independence • Musculoskeletal issues: Conditions like arthritis, scoliosis, back disorders, and bone weakness can also limit mobility and overall physical capacities
Sensory disabilities	These disabilities influence one or more of our senses—vision, hearing, touch, taste, or smell—and can be either congenital (present at birth) or acquired later in life Examples: • Visual impairments: These can range from low vision to total blindness, caused by conditions such as glaucoma, cataracts, and macular degeneration, highlighting the need for supportive environments • Hearing loss: This encompasses partial or total hearing impairment, with various causes, including trauma and genetics, emphasizing the importance of accessibility in communication • Deaf-blindness: This occurs when an individual experience both vision and hearing loss, as seen in conditions like Usher syndrome, pointing to the need for integrated support • Sensory processing disorder: Although not medically classified, this condition affects how individuals process sensory input and can greatly influence their daily experiences [6]
Intellectual disabilities	These neurodevelopmental disorders impact intellectual functioning and adaptive behavior, offering a unique perspective on learning and social interaction Examples: • Conditions like Down syndrome, Fragile X syndrome, and fetal alcohol spectrum disorders illustrate the diversity within this group • Individuals may face challenges in problem-solving, learning, social skills, and self-care, reminding us of the importance of tailored support • Approximately 1% of the population is affected, with about 85% experiencing mild symptoms, indicating that many individuals can lead fulfilling lives with appropriate support and accommodations [7]
Mental health disabilities	These disabilities encompass a range of disorders that influence emotional, psychological, and social well-being Examples: • Depression, anxiety, and bipolar disorder, etc. • The classification of the disability depends on its severity and impact on daily life • Approximately 1 in 8 people globally experience mental illness [8]

(continued)

Table 1 (continued)

Disability types	Description with examples
Neurological disabilities	These disabilities arise from conditions that affect the brain, spinal cord, or nerves Numerous conditions can cause these disabilities, such as: • Traumatic brain injury (TBI)- Resulting from a blow to the head • Stroke- caused by a disruption in blood flow to the brain • Neurodegenerative diseases- conditions like Alzheimer's and Parkinson's that lead to the gradual deterioration of nerve cells • Genetic disorders-disorders that affect the development or function of the nervous system • Infections- such as meningitis and encephalitis • Growth tumors- abnormal growths that can impact nerve function • By prioritizing early detection and rehabilitation, we can improve outcomes for individuals facing these challenges [9]
Chronic illness-related disabilities	Chronic illness-related disabilities emerge from long-term health conditions that can impact social, mental, and physical functioning Examples: • Conditions such as diabetes (leading to neuropathy and retinopathy), cancer (which may cause pain and fatigue), and cardiovascular diseases (like stroke and heart failure) highlight the need for comprehensive and integrative approaches to healthcare • By addressing these disabilities, we can enhance mobility, cognition, and independence, ultimately improving quality of life [10]
Communication disabilities	These disabilities affect an individual's ability to understand and produce speech and language • These challenges can stem from various causes, including autism, Down syndrome, intellectual disabilities, strokes, brain injuries, cleft lip/palate, and chemical exposure during pregnancy • By fostering inclusive communication strategies and providing necessary interventions, we can enhance communication skills, allowing individuals to connect more effectively with others and participate fully in their communities [11]

4.1 Accessibility Barrier

One of the most important challenges is the presence of accessibility barriers which encompasses physical, attitudinal and informational barriers.

Physical Barriers are a major challenge faced by disabled people. The inaccessible physical environment includes elements like poor design of public transport, inadequate sidewalks and pedestrian crossings, and construction of buildings that lack ramps, elevators and accessible restrooms. For example, a person with physical disability may struggle to access a public building without an elevator limiting them to access essential services [13, 14].

Attitudinal barriers like stigma and discrimination can have a major impact on the life of a disabled. Negative stereotypes and assumptions about their abilities may lead to social isolation, exclusion and limited opportunities. Employers may underestimate the capabilities of people with impairments which could lead to fewer job offers along with lack of awareness on how to accommodate impaired workers. A person with a visible impairment, for example, may have their work chances hampered by an employer's unfair assumption that they are incapable of carrying out specific job obligations. These lead to lower pay, greater unemployment rate and poor labour force participation when compared to persons without disabilities [15].

Informational barriers that include a lack of accessible information and communication further marginalize people with impairment. This includes public announcements that are incomprehensible to those with hearing impairments, documents in inaccessible formats (such as those without big print or audio versions), and websites that are inaccessible (such as those with inadequate keyboard navigation or no alt-text for images).

4.2 Financial Insecurity

Another significant obstacle is economic inferiority. Disability-related unemployment and underemployment rates are higher when discrimination occurs in the workplace. People with disabilities frequently receive lesser pay for equivalent work than their non-disabled counterparts, even when they are employed. All these factors contribute to low income and poverty. In addition to the costs associated with their impairment, people with disabilities frequently must pay for specialist equipment, home adaptations, and assistive technologies like wheelchairs and hearing aids. The high cost of assistive technologies, such as cochlear implants for those with hearing impairments or communication devices for those with communication impairments, prevents them from using them. They have trouble getting access to high-quality healthcare because of their financial limitations. For example, people with neurological disability often find difficulty in getting specialized care, diagnostic tools, and treatments as they are costly [13].

4.3 Social Exclusion

Misconceptions about disabilities, along with social stigma and low self-esteem, can create barriers for individuals with disabilities to participate fully in society. However, by addressing these misconceptions and fostering a more inclusive environment, we can enhance their involvement in public events, sports, recreational activities, and community gatherings. Improving accessibility infrastructure and providing robust support networks will significantly improve their quality of life. Moreover, individuals with intellectual disabilities can thrive when given the right support for their psychosocial development. By equipping them with tools to manage emotions and build relationships, we can help them connect more meaningfully with others. Combating feelings of isolation and misunderstanding is crucial; promoting awareness and acceptance can dismantle stigmas and discrimination surrounding their unique abilities. By challenging cultural taboos and fostering understanding, we create a more supportive society where individuals with mental disabilities feel empowered to seek help and share their experiences [13].

4.4 Educational Disparities

Disabled individuals often face significant challenges when trying to access inclusive education. These challenges include inaccessible physical environments, a shortage of trained teachers, and a lack of assistive technology. As a result, they may experience lower educational attainment and fewer employment opportunities. Some examples of physical barriers that students with disabilities encounter when seeking educational services include heavy doors, inaccessible restrooms, a lack of ramps or elevators in multi-story school buildings, and inadequate transportation options to and from school. Additionally, post-secondary students may struggle to find accessible student housing, which presents further difficulties [13].

Specialized formats, like Braille, sign language, or audio aids, are usually required for students with sensory disabilities to access educational resources, which are frequently insufficient. Individuals with intellectual disabilities have serious learning difficulties due to cognitive limitations like slower brain processing, short-term memory problems or trouble in grasping abstract ideas. Proper teaching methods and early interventions can successfully tackle this situation [13].

4.5 Healthcare Disparities

Disparities in healthcare are also common, with financial limitations, transportation issues, and a shortage of accessible healthcare providers making it difficult for individuals with disabilities to receive high-quality medical care. It was challenging for people with disabilities to effectively explain their issues due to the lack of time allocated for medical examinations and treatments. For instance, people with mental disabilities face the problem of highly expensive therapy, medicine, and support services, as well as the scarcity of mental health specialists and facilities, particularly in rural areas.

People with physical disabilities frequently encounter significant barriers to accessing healthcare services. For example, a woman with mobility impairment may be unable to undergo a crucial mammogram because the equipment requires her to stand, potentially delaying diagnosis and increasing health risks. Similarly, the lack of accessible weight scales, such as those that accommodate wheelchairs, can create obstacles in basic health assessments. Individuals with neurological disabilities face unique challenges, primarily stemming from the complex nature of neurological conditions. The intricate interplay of symptoms and the subtle variations between conditions often lead to misdiagnosis or delayed diagnosis, impacting the effectiveness of treatment and overall health outcomes [16].

Lack of access to vital support services frequently causes major challenges for people with communication problems. Their capacity to acquire and enhance communication skills may be hampered by the scarcity of Speech-Language Pathology (SLP) services. Additionally, their capacity to successfully express themselves may be significantly limited by a lack of training and access to Augmentative and Alternative Communication (AAC) tools. For instance, those with severe speech impairments could not have access to tools like voice synthesizers or communication boards, which would have a major negative influence on their quality of life, education, and social interaction [16].

Beyond the obstacles stated above, people with disabilities also have major obstacles in the areas of legal access and transportation. Social contact and independence can be severely restricted by inadequate public transit systems. For instance, a person's capacity to get to work, receive medical care, or engage in community activities may be severely limited by infrequent bus schedules, inconvenient routes, and a lack of wheelchair accessibility on public transportation. In addition, individuals with impairments frequently face substantial legal obstacles. These include communication challenges during court procedures, inaccessible courtrooms, and a shortage of attorneys with knowledge of disability rights. For instance, a person with a vision impairment would have trouble accessing legal documents if they are not available in accessible formats like braille or electronic text. Furthermore, disabled persons frequently have limited opportunities to meaningfully participate in policy-making processes, which makes it difficult for them to advocate for their own needs and help create inclusive policies [13].

Along with the above-mentioned challenges individuals with specific disabilities face unique challenges. People with sensory impairments, for example, frequently need help or direction to navigate physical environments and rely extensively on assistive equipment such as canes, screen readers, or cochlear implants. Because of their unpredictable symptoms and numerous medical appointments, people with chronic illnesses often struggle to maintain stable employment. People with invisible disabilities, such as chronic pain or mental health issues, have an especially big obstacle. They frequently encounter suspicion or disbelief from others since their ailments are not always obvious, which causes them to feel alone and frustrated. Fatigue and "brain fog" are symptoms that can make these difficulties worse and increase emotions of helplessness, loneliness, and rage [17].

Figure 3 A Venn diagram illustrates the connections and overlaps among challenges and causes of disabilities, highlighting how they are interrelated and influence one another. Table 2 shows the list of causes of disabilities used in the Venn diagram.

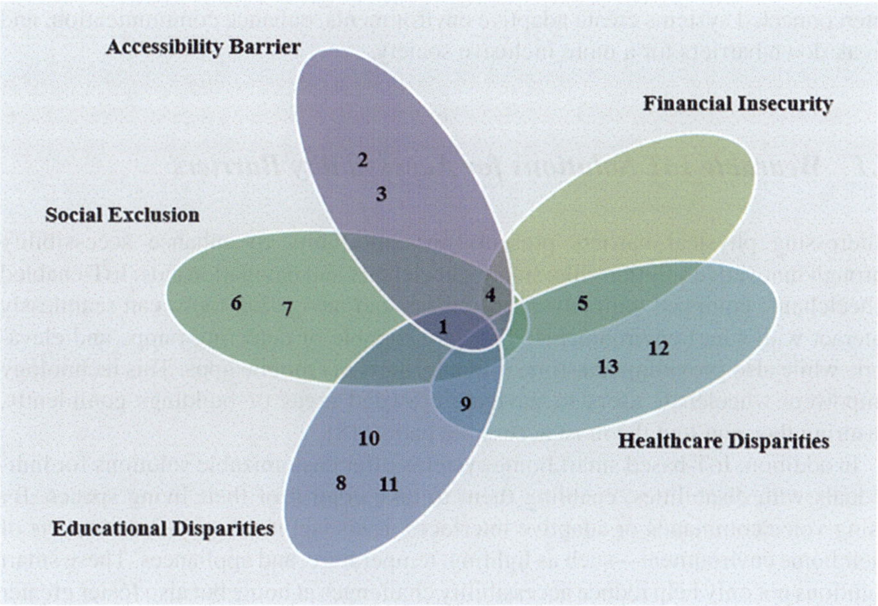

Fig. 3 A Venn diagram shows the challenges and causes of disabilities

Table 2 List of causes and their notations used in the Venn diagram

1	Lack of infrastructure
2	Social stigma
3	Informational barriers
4	Discrimination
5	High expense
6	Misconceptions
7	Lack of emotional support
8	Lack of affordable solutions
9	Inadequate transportation
10	Insufficient resources
11	Cognitive limitations
12	Inadequate equipment
13	Lack of support services

5 Wearable IoT Solutions for Various Disability Challenges

The wearable IoT is changing our world by offering improved connectivity and automation, especially helping people with disabilities by enhancing their accessibility, independence, and quality of life. Smart devices, wearable technologies, and interconnected systems create adaptive environments, enhance communication, and break down barriers for a more inclusive society.

5.1 Wearable IoT Solutions for Accessibility Barriers

Addressing physical barriers presents an opportunity to enhance accessibility through innovative solutions like smart wheelchairs and navigation aids. IoT-enabled wheelchairs, equipped with advanced sensors and navigation tools, can seamlessly interact with smart environments. They are capable of detecting ramps and elevators, while also providing real-time obstacle alerts via mobile apps. This technology empowers wheelchair users to navigate crowded areas or buildings confidently, ensuring they can find the most accessible paths [18].

In addition, IoT-based smart home systems offer customizable solutions for individuals with disabilities, enabling them to take control of their living spaces. By using voice commands or adaptive interfaces, users can manage various aspects of their home environment—such as lighting, temperature, and appliances. These smart solutions not only help reduce accessibility challenges at home but also foster greater independence in daily routines [5].

In addition to homes, IoT-based building accessibility systems and public spaces are becoming part of modernization. Smart elevators, for instance, allow doors to be controlled by voice commands or mobile applications. Additionally, exoskeletons and IoT-enabled smart canes (such as the WeWalk Smart Cane) help individuals with spinal injuries or visual impairments. All these prove that wearable IoT can change the physical accessibility of disabled people [19].

Addressing attitudinal barriers like stigma and discrimination against individuals with disabilities is essential for creating an inclusive atmosphere everywhere. Integrating Internet of Things (IoT) technologies can significantly reduce stigmatization and enhance self-esteem among individuals with disabilities. Wearable devices, such as smartwatches equipped with specialized applications for emotional regulation, have been developed to help young adults who have intellectual and developmental disabilities manage daily stressors, thereby facilitating smoother social interactions [20].

In workplaces, wearable IoT technologies can significantly help by enhancing accessibility, raising awareness, and promoting equitable opportunities. IoT-enabled assistive devices like communication or mobility aids help employees to engage fully in workplace activities [21]. Using IoT technologies to enable remote work options can greatly benefit individuals with disabilities. By allowing them to work in environments that are customized to their specific needs, workplace discrimination can be reduced, and employment opportunities can be expanded. For instance, telepresence robots enable individuals with mobility impairments to participate in on-site meetings virtually, offering a seamless way to be present without physical limitations [22].

Addressing informational barriers is essential to ensure inclusivity for individuals with impairments. Wearable IoT technologies offer innovative solutions to mitigate this challenge. For example, assistive listening devices and real-time transcription services help to understand public announcements [23].

IoT-enabled Sign Language Interpreters and Inclusive Digital Platforms can help improve the information accessibility of disabled individuals. So, all these together improve their standard of living with the help of Wearable IoT.

5.2 *Wearable IoT Solutions for Financial Insecurity*

Wearable IoTs can help reduce economic challenges for individuals with disabilities by improving employment opportunities, increasing workplace inclusivity, lowering healthcare costs, and providing affordable assistive technologies.

Wearable IoTs enhance employment opportunities for individuals with disabilities by offering remote work enablers like VR headsets and haptic feedback gloves, real-time skill development tools like Muse Headbands and AR-based apps, and workplace accessibility solutions like IoT integrated speech-to-text processors such as Dragon NaturallySpeaking. Another instance is for visually impaired individuals using OrCam MyEye and smart wearables, to participate equally in professional activities [24].

Wearable IoTs help reduce the costs of advanced assistive technologies by benefiting from economies of scale and IoT-driven resource sharing. Devices like IoT-powered hearing aids receive cloud-based updates, lowering manufacturing and maintenance costs, while online platforms enable sharing or renting assistive devices, making them more accessible to low-income users [25]. For example, platforms like Wheel Share use IoT to manage shared wheelchair stations, minimizing individual costs and maximizing availability.

IoT systems are transforming access to high-quality services by making them more affordable. IoT-supported insurance models use personalized data from wearables to offer premium discounts, which reduce financial burdens for individuals with disabilities. Smart rehabilitation centres, equipped with IoT-based devices, provide affordable rehabilitation services like virtual physiotherapy sessions and gait analysis [26]. For example, IoT-powered platforms like Virtual Health Assistant by Babylon offer affordable consultations, saving on diagnostic costs for users who need frequent specialist visits.

In addition, IoT-powered devices like smart stethoscopes (e.g., Eko DUO) and blood pressure cuffs connect to telemedicine platforms, reducing the need for costly in-person visits to specialists. Medication management systems, such as IoT-enabled pillboxes and trackers, help users adhere to prescribed treatments, preventing costly health deterioration [27]. For example, individuals with neurological disabilities may use EEG-based wearables like NeuroPace, connected to healthcare apps, reducing the need for frequent and expensive clinic visits. These innovations make healthcare more efficient and affordable for individuals with disabilities by enabling early intervention, remote care, and better medication adherence.

5.3 Wearable IoT Solutions for Social Exclusion

Wearable IoTs present promising solutions to the challenges of social exclusion experienced by individuals with disabilities. By fostering a sense of inclusion and promoting self-reliance, these technologies enhance communication and engagement, helping to break down stigmas. As a result, they contribute to improved psychosocial development and support the social integration of individuals, creating a more inclusive society for everyone.

Wearable IoTs enable individuals with disabilities to actively participate in public life, events, and recreational activities by using social connectivity devices like IoT-enabled smartwatches, which facilitate communication through text, voice, and sign language translation, allowing users to interact confidently. AR/VR wearables provide immersive experiences that help individuals participate virtually in public events, sports, or community gatherings if they face physical barriers. IoT-based navigation assistants, such as the WeWalk Smart Cane, equipped with GPS, obstacle detection, and public transport integration, help visually impaired individuals independently explore social spaces [5]. For example, a person with hearing impairment can use real-time captioning through glasses connected to IoT, enabling them to follow conversations during social gatherings.

IoT-powered speech-assistive technologies, such as speech-to-text and text-to-speech conversion tools, allow individuals with communication impairments to express themselves more fluently in group settings, effectively challenging stereotypes of incapability. Additionally, wearable AI companions that integrate conversational artificial intelligence with IoT can simulate meaningful interactions, reducing feelings of isolation and promoting social learning [28]. These devices offer users opportunities to engage in conversations, enhancing their social skills and self-confidence.

For example, a person with autism spectrum disorder can use IoT-linked emotional coaching devices to recognize stress indicators and receive real-time guidance during social interactions, improving their engagement and reducing anxiety. Another instance, AR glasses, such as OrCam MyEye, assist individuals with visual or cognitive impairments by reading texts aloud and identifying faces in real-time [29]. Collectively, these IoT solutions play a pivotal role in promoting self-confidence and addressing misconceptions about disabilities, contributing to a more inclusive society.

Wearable IoT technologies support the psychosocial development of individuals with intellectual disabilities by addressing emotional, social, and interpersonal challenges. IoT-based social skills training platforms, such as VR exercise programs, offer safe environments for practicing interpersonal interactions. Collaborative gaming wearables promote teamwork and social engagement through interactive play [30]. IoT-enabled behavioural insights from smart devices equipped with biosensors provide valuable data to caregivers for timely interventions. Wearable stress sensors and devices that detect emotional states help caregivers understand and support a child's emotional development [31]. Overall, these Wearable IoT solutions enhance emotional regulation, social skills training, and behavioral insights, contributing to the well-being and integration of individuals with intellectual disabilities.

5.4 Wearable IoT Solutions for Educational Disparities

Specialized Formats for Sensory Disabilities: Wearable IoT empower sensory-impaired students through adaptive technologies that enhance accessibility and participation in education. Real-time sign language interpretation using IoT-powered gloves, such as SignAloud, converts sign language gestures into text or audio, enabling seamless communication for deaf students with teachers and peers [32]. Like the Braille eBooks Dot Watch, text-to-Braille translators dynamically transform standard text into Braille outputs, providing an accessible format for visually impaired learners [33]. IoT-based audio resources, such as smart glasses with integrated audio narration (e.g., OrCam MyEye), facilitate the comprehension of visual materials by offering real-time audio descriptions. For instance, a visually impaired student using Google Glass with OCR technology can use here to access printed materials through text-to-speech functionalities.

In the case of Access to Inclusive Housing, finding accessible accommodations can be a significant challenge for graduate students with disabilities, which is addressed effectively by Wearable IoT solutions. IoT-driven housing management systems, such as the My Access Housing app, utilize sensors to help students identify, book, and manage accommodations with smart features like voice-activated devices, automated lighting, and accessible kitchen appliances. IoT home safety wearables, like Smart Locks with accessibility options, ensure secure and seamless access for residents.

Cognitive Support for Intellectual Disabilities: Wearable IoTs significantly enhance the learning experience for students with intellectual disabilities by addressing their unique cognitive challenges. IoT-based learning apps, such as Special Learner, engage students through interactive, multisensory content, ensuring sustained focus and effective learning. Memory aids like the Emotiv Insight wearable detect attention levels and help break down complex subjects into manageable modules tailored to individual learning speeds. IoT sensors and wearables also track classroom behavior in real-time, alerting teachers to adaptive teaching strategies. Devices like Muse use EEG sensors to provide insights into student focus, improving classroom learning outcomes [34].

Enhancing Teacher Training and Interventions: IoT wearables support educators by enabling them to adapt to the specific needs of students with disabilities. IoT teacher dashboards paired with wearables offer real-time insights into student engagement, progress, and challenges, facilitating dynamic and individualized lesson plans. Virtual reality (VR) headsets, such as Oculus, simulate inclusive classroom scenarios, allowing teachers to train in accommodating diverse learning needs effectively. Assistive communication wearables, like smart pens and earpieces, enable teachers to create interactive lessons tailored for sensory and intellectual disabilities [35]. For instance, a teacher wearing IoT-enabled Google Glass can interact with deaf students by displaying captions of their speech in real-time, improving communication and inclusion.

5.5 Wearable IoT Solutions for Healthcare Disparities

Wearable IoT solutions transform healthcare for individuals with disabilities by enhancing accessibility, communication, and diagnostic capabilities.

Addressing Financial Limitations and Accessibility: Wearable IoT devices, such as health monitors, allow for the remote tracking of vital signs. This reduces the need for frequent in-person medical visits, ultimately lowering healthcare costs. Additionally, IoT-enabled assistive devices and applications enhance independence and provide better access to essential services for individuals with disabilities. Wearable health monitors like Fitbit and Biovitals track vital signs in real-time, helping those with chronic or neurological conditions identify potential health issues early, which can prevent costly emergency interventions.

Improving Transportation and Physical Barriers in Healthcare Access: IoT technology enhances mobility and accessibility for individuals with disabilities. Smart devices, including IoT-connected wheelchairs, help users navigate physical spaces more effectively. Moreover, IoT-enabled communication devices improve interaction and accessibility in various environments [36].

Improving Diagnosis and Treatment for Neurological Disabilities: IoT-assisted wearable sensor systems are vital in healthcare, enabling real-time monitoring and support for individuals with neurological conditions. These systems enhance treatment efficiency and improve the quality of medical services delivered [37].

Supporting Communication Disabilities: Wearable IoT devices aid communication for individuals with hearing impairments. For example, wearable haptic assistive devices have been developed to facilitate communication with hard-of-hearing individuals [38].

Enhancing Mental Health Accessibility: IoT-based healthcare monitoring systems contribute to improved quality of life by enabling continuous health monitoring and timely interventions, which are essential for mental health care.

These advancements demonstrate the significant impact of Wearable IoT solutions in bridging healthcare disparities for people with disabilities.

5.6 Wearable IoT Solutions for Addressing Legal and Transportation Barriers

Wearable IoT technologies are transforming accessible transportation by enhancing mobility and independence for individuals with disabilities. IoT-integrated public transit systems provide real-time updates on accessible features like ramps and elevators, improving convenience. Connected ride-sharing platforms, such as Uber WAV, match users with accessible vehicles while wearables communicate special needs to drivers. Personal navigation devices like Microsoft's Soundscape use GPS and spatial audio to assist visually impaired users, and smart infrastructure, such as Accessible Pedestrian Systems (APS), enhances urban mobility by signalling when it is safe to cross streets. These innovations showcase how Wearable IoT solutions address the transportation challenges faced by people with disabilities [39].

Wearable IoT technologies are enhancing legal accessibility by improving communication, document access, and participation in legal proceedings for individuals with disabilities. IoT-enabled court accessibility tools, such as real-time transcription devices, provide instant captions during hearings, aiding those with hearing impairments. Smart glasses equipped with text-to-speech capabilities assist visually impaired individuals in reading legal documents and navigating courtrooms independently. Cloud-connected devices facilitate access to legal documents in various accessible formats, ensuring timely and private retrieval of information. Additionally, IoT-facilitated remote participation tools enable individuals with disabilities to engage in legal proceedings from their homes, overcoming physical and transportation barriers. These advancements demonstrate the potential of Wearable IoT solutions to make legal systems more inclusive and accessible [36].

6 Limitations of Wearable IoT

Wearable IoT devices provide significant benefits to individuals with disabilities, but they also face several limitations that impact their effectiveness. These limitations can be categorized into technical, ethical, and implementation challenges.

(a) Technical issues:

Devices like smart canes and haptic feedback tools have the potential to greatly assist users, but they rely on stable internet connectivity and accurate sensors. Improving connectivity options in areas with limited service can enhance user experience. Additionally, extending battery life in wearables would allow for more convenient, long-term use. Further, by addressing the high initial costs of advanced devices like OrCam MyEye and WeWALK, we can make these important tools more accessible to low-income individuals.

(b) Ethical issues:

As we advance with wearable technology, it's essential to address data privacy risks associated with the continuous tracking of user locations. Implementing stronger safeguards would help protect personal security. Organizations should also prioritize compliance with data regulations, ensuring that systems are robust enough to uphold standards like GDPR and HIPAA. By fostering industry-wide adherence to these regulations, we can build greater trust among users. Additionally, focusing on the development of more personalized wearable devices can improve accuracy in functionality, allowing for better support that accommodates the fluctuating symptoms and needs of individuals with chronic conditions.

The continuous use of wearables can lead to physical discomfort, particularly during flare-ups or long-term illnesses, which adds to the challenges faced by users. The subjective nature of experiences, such as emotional struggles or pain, is often difficult to quantify accurately with devices like Mood Beam. Over-reliance on algorithms may result in misinterpretations or the omission of subtle indicators that signal worsening conditions. Additionally, many wearable IoT tools lack medical-grade certification, which limits their acceptance in healthcare interventions. The flood of data and constant updates from these devices can contribute to cognitive overload, rather than alleviating it.

(c) Implementation challenges:

Technical and accessibility barriers continue to exist, particularly in rural or under-resourced areas with inadequate infrastructure. This situation limits the effectiveness of advanced IoT healthcare devices, such as connected mammography systems and IoT stethoscopes. The high costs associated with these tools further complicate the issue, as even basic healthcare IoT solutions can be prohibitively expensive, making them inaccessible for many individuals with disabilities. Additionally, operating advanced devices and interpreting their data require specialized training; however, there is a shortage of skilled professionals in this field.

For example, the incomplete integration of IoT communication devices into legal systems, such as those used in courtrooms, restricts their effectiveness in ensuring access. This highlights the need for broader policy enforcement to promote universal adoption. Similarly, in the transportation sector, smart public transport IoT systems depend on government implementation. Consequently, cities with low technology adoption experience heightened accessibility challenges. Furthermore, the lack of infrastructure development exacerbates these financial difficulties [40].

7 Conclusions

The integration of wearable IoT technology in healthcare is a powerful opportunity to transform the lives of individuals with disabilities. This technology not only enhances their quality of life but also increases their participation in the labor market, creating a meaningful impact. Moreover, the continuous tracking of patient data empowers stakeholders with real-time information, leading to significant reductions in medical costs and errors. It optimizes resource utilization and ensures well-coordinated medical services, guaranteeing that patients receive timely and comprehensive care.

To fully harness the advantages of wearable IoT devices for individuals with disabilities, we must proactively tackle the challenges they face. We need to focus on critical areas such as improving data security, enhancing interoperability, overcoming technical hurdles, and addressing ethical considerations. While the diversity of devices and stakeholders may introduce complexity, it also presents valuable opportunities for collaboration across the healthcare ecosystem. By prioritizing the standardization of devices and clearly defining stakeholder roles, we will strengthen data security and effectively address technical and ethical challenges.

Looking ahead, we must capitalize on the tremendous potential for future research to integrate machine learning techniques, innovative data analytics software, and blockchain technology within wearable IoT solutions. Cross-sector collaboration among policymakers, academics, industry leaders, and service providers is essential to develop sustainable and impactful healthcare applications. By working together, we will achieve better outcomes for individuals with disabilities and significantly enhance the overall effectiveness of these technologies. This collaborative spirit is key to driving innovation and effecting positive change in healthcare.

References

1. Li, C., Wang, J., Wang, S., Zhang, Y.: A review of IoT applications in healthcare. Neurocomputing **565**, 127017 (2024)
2. Xu, X., Fu, C., Camacho, D.: Internet of things for emotion care: advances, applications, and challenges. Cogn. Comput. **16**, 2812–2832 (2024)
3. Semary, H.E., Al-Karawi, K.A., Abdelwahab, M.M.: A review on internet of things (IoT)-related disabilities and their implications. J. Disabil. Res. **3**(2), (2024)
4. Four Most Common Types of Sensory Disabilities.: Enable Me. https://www.enableme.ke/en/article/4-most-common-types-of-sensory-disabilities-1535#:~:text=Sensory%20disabilities%20are%20the%20disabilities,others%20are%20long%2Dlife%20disorders. Accessed 14 Jan 2025
5. Luna, R.E.: What is Intellectual Disability? American Psychiatric Association. https://www.psychiatry.org/patients-families/intellectual-disability/what-is-intellectual-disability#:~:text=Additionally%2C%20the%20intellectual%20and%20adaptive,age%2018%20years%20for%20diagnosis. Accessed 02 Jan 2025
6. Laderer, A.: Is Mental Illness a Disability? Charlie Health. https://www.charliehealth.com/post/is-mental-health-a-disability#:~:text=Mental%20health%20conditions%20can%20indeed,activities%2C%20carry%20out%20tasks%20at (2023) Accessed 13 Nov 2024
7. Hirtz, D., Thurman, D.J., Gwinn-Hardy, K., Mohamed, M., Chaudhuri, A.R., Zalutsky, R.: How common are the common neurologic disorders? Neurology **68**(5), 326–337 (2007)
8. Chou, C.Y., Chiu, C.J., Chang, C.M.: Disease-related disability burden: a comparison of seven chronic conditions in middle-aged and older adults. BMC Geriatr. **21**, 201 (2021)
9. Tee-Melegretto, R.A.: What to Know About Communication Disorders. MedicalNews Today. https://www.medicalnewstoday.com/articles/communication-disorders. Accessed 10 Dec 2024
10. Deng, L., Chen, J., Li, D.: Factors influencing intentions of people with hearing impairments to use augmented reality glasses as hearing aids. Behav. Sci. **14**(8), 728 (2024)

11. World Health Organization Report.: https://www.who.int/campaigns/international-day-of-persons-with-disabilities/2024 (2024). Accessed 05 Nov 2024
12. Gadzhimusieva, D., Gorelova, A., Meliá, S., Lledó, G.: Enhancing accessibility in academic buildings: a discrete event simulation approach for robotic assistance. IEEE Access **12**, 126885–126898 (2024)
13. Jense, J., Sathiyandra, S., Rochford, M.: Work participation among people with disabilities: does the type of disability influence the outcome? Soc. Policy J. NZ. 24 (2005)
14. Moon, N.W., Baker, P., Goughnour, K.: Designing wearable technologies for users with disabilities: accessibility, usability, and connectivity factors. J. Rehabil. Assistive Technol. Eng. **6** (2019)
15. Disability Barriers to Inclusion. Disability Inclusion Disability Barriers to Inclusion|Disability Inclusion|CDC. Accessed 12 Dec 2024
16. Burke, A.C, Cronkwright, S.: The Challenges of Living with Invisible Illness, Psychology Today. https://www.psychologytoday.com/ca/blog/stroke-of-awareness/202104/the-challenges-of-living-with-an-invisible-illness. Accessed 12 Dec 2024
17. Sonekar, M.M., Waghulde, K.D., Patil, L.N.: Autonomous navigated IOT enabled smart wheel chair—a healthcare and mobility solution. J. Electr. Syst. **20**(2), 709–720 (2024)
18. Ghasemi, P., Mirabi, M.: Smart assistive devices: a survey of IoT solutions for physical accessibility. IEEE Sens. J. **20**(1), 24–32 (2018)
19. McGinnis, R.: New Smartwatch Design Will Benefit Young Adults with Disabilities-2020. College of Engineering and Computing. George Mason University. https://cec.gmu.edu/news/2020-02/new-smartwatch-design-will-benefit-young-adults-disabilities
20. Kruse, D., Schur, L., Johnson-Marcus, H.A., Gilbert, L., Di Lallo, A., Gao, W., Su, H.: Assistive technology's potential to improve employment of people with disabilities. J. Occup. Rehabil. **34**(2), 299–315 (2025)
21. Takeuchi, K., Yamazaki, Y., Yoshifuji, K.: Avatar work: telework for disabled people unable to go outside by using avatar robots. In: Companion of the 2020 ACM/IEEE International Conference on Human-Robot Interaction (HRI'20), pp. 53–60. Association for Computing Machinery, New York, NY, USA
22. Adasitecompliance.: Listening Devices Improving Audio Accessibility in Public Spaces. https://adasitecompliance.com/assistive-listening-devices-improving-audio-accessibility-public-spaces/. Accessed 19 Oct 2024
23. Orcam.: Personal Stories: How AI and IoT Are Transforming Assistive Tech and Changing Lives. https://www.orcam.com/en-us/blog/personal-stories-ai-iot-transforming-assistive-tech-changing-lives (2024). Accessed 12 Nov 2024
24. Jia, X., Guo, Z., Wang, H.: Assistive IoT devices and their economic impacts on accessibility. J. Econ. Perspectives Technol (2019)
25. Kelly, J.T., Campbell, K., Gong, E., Scuffham, P.: The internet of things: impact and implications for health care delivery. J. Med. Internet Res. **22**(11), e20135 (2020)
26. Monge, J., Ribeiro, G., Raimundo, A., Postolache, O., Santos, J.: AI-based smart sensing and AR for gait rehabilitation assessment. Information **14**(7), 355(2023)
27. Kalantari, N., Zheng, H., Graff, H., Evmenova, A., Motti, V.: Emotion regulation for neurodiversity through wearable technology. 2021 9th International Conference on Affective Computing and Intelligent Interaction (ACII), Virtual Event, 28 September–01 October 2021 (2021)
28. Nguyen, X.T., Koopman, J., van Genderen, M.M., Stam, H.L.M., Boon, C.J.F.: Artificial vision: the effectiveness of the Orcam in patients with advanced inherited retinal dystrophies. Acta Ophthalmol. **100**(4), e986–e993 (2022)
29. Greuter, S., Watson, J., Balandin, S.: Social adventure: designing interactive smart speaker social skills games for people with intellectual disabilities. Int. J. Gaming Comput.-Mediated Simul. **14**, 1–21 (2022)
30. Park, S.B., Ju, Y., Kwon, H., Youm, H., Kim, M.J., Chung, J.: Effect of a cognitive function and social skills-based digital exercise therapy using IoT on motor coordination in children with intellectual and developmental disability. Int. J. Environ. Res. Public Health **19**(24), 16499 (2022)

31. Chin, M.: Wearable-Tech Glove Translates Sign Language Into Speech In Real Time. https://newsroom.ucla.edu/releases/glove-translates-sign-language-to-speech. Accessed 26 Jan 2025
32. D'Mello, G.: Six MIT Students Built A Pocketable Text-To-Braille Translator That Will Truly Empower Blind People. IndiaTimes. https://www.indiatimes.com/technology/gadgets/mit-students-built-a-text-to-braille-translator-that-could-give-the-vision-impaired-more-independence-321339.html. Accessed 17 Jan 2025
33. Austin, A.: 8 Helpful Apps for Students with Special Needs. Connections Academy. https://www.connectionsacademy.com/support/resources/article/apps-for-students-with-special-needs/ (2022)
34. A Guide to IoT Use Cases in Education: Unlocking the Future of Learning. Matellio https://www.matellio.com/blog/iot-use-cases-in-education/ (2023)
35. Roussey, B.: How Internet of Things (IoT) Devices Can Improve Accessibility. Accessibility.com. https://www.accessibility.com/blog/how-internet-of-things-iot-devices-can-improve-accessibility. Accessed 10 Dec 2024
36. Mamdiwar, S.D.R.A., Shakruwala, Z., Chadha, U., Srinivasan, K., Chang, C.Y.: Recent advances on IoT-assisted wearable sensor systems for healthcare monitoring. Biosensors **11**(10), 372 (2021)
37. Goyal, M., Basavarajappa, G.: A wearable IoT based assistive device to aid communication with hearing impaired. 2023 IEEE Microwaves, Antennas, and Propagation Conference (MAPCON), Ahmedabad, India, 2023, pp. 1–4 (2023)
38. Mazur, B.: Iot In Public Transportation: How Innovation Is Transforming Urban Mobility. Ignitec. https://www.ignitec.com/insights/iot-in-public-transportation-how-innovation-is-transforming-urban-mobility/. Accessed 02 Jan 2025
39. Erdem, A., Eksin, E., Senturk, H., Yildiz, E., Maral, M.: Recent developments in wearable biosensors for healthcare and biomedical applications. TrAC, Trends Anal. Chem. **171**, 117510 (2024)
40. The UNDRR report. https://www.undrr.org/report/2023-gobal-survey-report-on-persons-with-disabilities-and-disasters (2023). Accessed 15 Nov 2024

Dr. Md. Anisul Islam is a Tenure-Track Assistant Professor in the Department of Management Science at the Shannon School of Business at Cape Breton University in Sydney, Nova Scotia, Canada. In his previous roles, he has made impactful contributions as a postdoctoral fellow on a significant project funded by the Canadian Institutes of Health Research (CIHR) and as a sessional instructor at the University of Manitoba. His research focuses on innovative solutions in healthcare, the Internet of Things, and blockchain technology, utilizing metaheuristics and machine learning methods to enhance sustainable supply chain networks. His extensive work has been published in prestigious research journals and presented at various international conferences, thus playing a vital role in advancing knowledge and driving innovation in these important fields.

Radhika Ramachandran Nair is a postgraduate student in Business Analytics at the Shannon School of Business, Cape Breton University, Sydney, Canada. She holds an undergraduate degree in Computer Engineering from Mahatma Gandhi University, Kerala, India. Her research interests include Decision Intelligence, AI Solutions, Data Governance, and Autonomous Diagnostics. She has actively participated in various research projects related to Data Analytics, Automation, and IT Solutions.

Gadha Jyothi Prasad is a Postgraduate Business Analytics student at the Shannon School of Business at Cape Breton University, Sydney, Canada. She received her Master's in Mathematics from Mar Ivanios College, Kerala, India. Her research interests mainly lie in Graph Theory, linear algebra, Operations Research, AI, and Machine Learning. She has presented research papers in national seminars and contributed to various projects integrating mathematics, healthcare, and analytics.

Dr. Rafiqul Chowdhury received his Ph.D. in Biostatistics from the Institute of Statistical Research and Training (ISRT), Dhaka University, Bangladesh. After that, he completed a post-doctoral research fellowship in the Department of Mathematics and Statistics of the University of New Brunswick, Canada. He is an assistant professor at the Shannon School of Business at Cape Breton University, Sydney, NS, Canada. Previously, he taught at the Department of Mathematics and Statistics, University of Fraser Valley (UFV), Abbotsford, BC; School of Mathematical and Computational Sciences (SMCS), University of Prince Edward Island (UPEI), Charlottetown, PE; Department of Health Information Administration (HIA), FAHSN, Kuwait University, Kuwait and Computer and Information Technology Institute (currently: Institute of Information Technology), Jahangirnagar University, Savar, Dhaka, Bangladesh. His current research interest is developing statistical and machine learning models/frameworks to analyse big data and trajectory risk prediction collected longitudinally. His research interests include covariatedependent Markov models, repeated measure data, joint modeling, and models for big data analysis, among others. His recent and complete list of research can be seen on ResearchGate (Rafiqul Chowdhury).